Chances Are

Chances Are

The Only Statistics Book You'll Ever Need

Steve Slavin

MADISON BOOKS
Lanham • New York • Oxford

Published by Madison Books
4720 Boston Way
Lanham, Maryland 20706

12 Hid's Copse Road
Cumnor Hill, Oxford OX2 9JJ, England

Distributed by National Book Network

Library of Congress Cataloging-in-Publication Data

Slavin, Steve.
    Chances are : the only statistics book you'll ever need / Steve
Slavin.
      p. cm.
    ISBN 1-56833-107-X (cloth : alk. paper). — ISBN 1-56833-108-8
(pbk : alk. paper)
    1. Statistics. I. Title.
QA276.12.S573 1998
519.5—dc21   98-14496
                                                                  CIP

⊖™ The paper used in this publication meets the minimum requirements of
American National Standard for Information Sciences—Permanence of
Paper for Printed Library Materials, ANSI Z39.48–1984.
Manufactured in the United States of America.

Contents

Acknowledgments

I want to thank my editor, Kermit Hummell, for making this book possible, and Nancy Ulrich, the acquisitions editor, who negotiated a very fair contract.

Hazel Staloff, who typed my first book 15 years ago (also published by University Press of America), literally turned a rough manuscript into the book you are reading.

The copy editor, Tracy Davis, who caught and corrected my grammatical and mathematical errors, and Lynn Weber, the production editor, who coordinated the production process and kept all of us right on schedule.

I also want to thank my father, Jack, a retired high school math teacher, for setting the standard at a time when there actually *were* standards. And finally, I want to thank my family and friends whose advice and encouragement made my job a lot easier.

How to Use This Book

Statistics is a highly technical subject, replete with complex formulas and equations that seem written almost entirely in Greek. A basic course in statistics is required for most undergraduate majors in the social, natural, and physical sciences, not to mention those in education and business. Given the poor math backgrounds that most students bring to these courses, very few who complete statistics can apply any of the basic concepts that were covered.

At some time in our lives, nearly all of us will encounter statistics at school, on the job, or just reading a newspaper. Many job descriptions call for quantitative skills, or, less formally, the ability to crunch numbers. What they really want is someone who can work with numbers, draw graphs, and do simple statistical analysis. Working your way through this book will not make you into an instant statistician, but you certainly will qualify as a number cruncher.

You don't know it yet, but you really lucked out by choosing this book. My approach is to make every problem as simple as possible. And rather than placing the answers to every other problem at the end of the book, I present a full solution immediately after every problem.

You will be learning statistics by doing statistics. By the time you have worked your way through this book you will have done hundreds of problems. Statistics can appear quite intimidating because of all its unfamiliar terms and symbols as well as its rather complex formulas. Indeed, virtually every statistics book appears to be written largely in Greek. The only Greek letter you'll need to learn in this book is Σ, or sigma, which means "sum of." For example, ΣX means the sum of the X's. And $(\Sigma Y)^2$ means the sum of the Y's squared.

Do you need to know a lot of math to do the problems in this book? A smattering of very elementary algebra would be helpful. Mainly we'll be plugging numbers into formulas and solving them arithmetically.

What if your math skills are rusty? Well, if they're *really* rusty, I would recommend that you read *All the Math You'll Ever Need* (published by John Wiley and Sons), which I happened to write. You can work your way through this book in about three weeks to relearn any of the elementary, middle, and high school math you've forgotten.

After a few chapters you'll begin running into problems that call for doing a lot of multiplying and dividing. May you use a calculator? Definitely! In fact, it's extremely hard to do statistics without one. Is there anything else you'll need to do the problems in this book? Just one or two packages of graph paper and a ruler.

Will you have to memorize a lot of formulas? No! Perhaps the only formula you'll need to know is one that you probably have been using for years: the formula for computing the average, or mean:

$$\overline{X} = \frac{\Sigma x}{n}$$

The term $\overline{X}$ is the mean, or average, of X. The letter n stands for the number of terms; in this case, the number of X's. Putting it all together, we have: The mean of X is equal to the sum of X's divided by the number of X's. For example, what is the mean of 2, 5, and 8?

$$\overline{X} = \frac{\Sigma x}{n} = \frac{2+5+8}{3} = \frac{15}{3} = 5$$

You'll see that there are chapter reviews at the end of some chapters. These are the chapters that present a few different types of problems, so the problems in the chapter review serve as an end-of-chapter test. Please don't go on to the next chapter until you are confident that you have mastered all the material. Unless you already know a substantial amount of statistics and are using this book just to refresh your memory and sharpen your skills, you'll need to take your time working your way through this book. Statistics is hard enough as it is, especially if you are learning it on your own.

There are times when you'll run up against some very difficult problems. Don't get discouraged. Sometimes the best course of action is to put the book aside for an hour or two, or even for a couple of days, and then come back to it. But what if you *still* don't get it? Then I'd advise you to go on to the next topic, or even the next chapter. Almost no one who takes a statistics course actually learns every single thing that the course covers. So don't get bogged down in the stuff that you *can't* do. At some point you'll need to cut your losses and go on to the stuff you *can* do.

You'll find very little discussion of statistical concepts and virtually no statistical theory here. Basically I'll be showing you how to do particular types of problems step by step. Then I'll ask you to work out exactly the same types of problems. Once you've done that, you'll be able to check your work with my own step-by-step solutions. As I said before, you learn statistics by doing statistics.

Chapter 1

Presentation of Statistical Data

There are three main ways to present statistical data—in tables, in graphs or charts, and in words. Let's start with words.

When Lisa was five years old she weighed 42 pounds. Her weight rose to 49 pounds when she was six, 57 pounds when she was seven, 66 pounds when she was eight, 76 pounds when she was nine, and 87 pounds when she was ten. Imagine if we recorded her weight for the rest of her life!

There's got to be a better way. Actually there are *two* better ways—constructing a table or drawing a graph. We'll be doing both over the next few pages.

We'll work with tables and then move on to graphs and pie charts (which look exactly like pies, especially pizza pies). Tables are a good way to present data in an orderly, easy to read way. Of course if these tables contain hundreds of numbers, they may not be too easy to read. Luckily the tables you'll be working with right now are quite simple.

Face it, simple or complex, tables are boring. We can take the same data and make it into a much more interesting graph. And if we happen to have a set of magic markers, or better yet, a color printer, then there's almost no limit to what we can draw. Except, of course, the data we're working with.

Simple Tables

The U.S. government, economics textbook writers, and others with a whole lot on their minds are great at compiling tables containing hundreds of numbers. While these tables certainly squeeze a great deal of data into a small amount of space, they are often very difficult to use. We're going to start out with very simple, easy-to-read tables and work our way up to tables that contain a little more data.

What's data? Data (which may be used as both a singular and a plural noun) is a collection of factual information, usually statistics or numbers. In this book, when we talk about data, we're talking about numbers. Right now we're going to be compiling tables containing a few numbers.

In Table 1.1 we show Lisa's weight each year from the age of five to the age of ten. This is as easy as it gets.

Table 1.1: Lisa's Weight from Age Five to Age Ten

Age	Pounds
5	42
6	49
7	57
8	66
9	76
10	87

Now let's add a new set of data. Let's add Karen's weights at the same ages. That's done in Table 1.2.

Table 1.2: Lisa's and Karen's Weight from Age Five to Age Ten

Age	Lisa's Weight in Pounds	Karen's Weight in Pounds
5	42	39
6	49	45
7	57	51
8	66	58
9	76	66
10	87	75

Sometimes we use tables to make comparisons. Suppose we wanted to compare the number of seats in the House of Representatives held by Democrats and Republicans in each of the four regions of the country. We've done that in Table 1.3, using the results of the election of 2020. Remember, you saw it first right here.

Table 1.3: House Seats Won by Democrats and Republicans in the Election of 2020, by Region

Region	Democrats	Republicans	Total
East	64	39	103
South	49	69	118
Midwest	54	47	101
West	60	53	113
	227	208	435

I've been doing all the work so far. Now I'd like *you* to draw up a table. I'll give you all the data and we'll see what you can do with it. We'll start you off with an easy one.

The population of Upper Volta was 1,206,198 in 1920. It rose to 1,372,061 in 1930. It fell to 1,338,926 in 1940. Then it rose to 1,404,228 in 1950. And it rose again to 1,536,490 in 1960. Compile a table of Upper Volta's population from 1920 to 1960. Do it on a separate piece of paper. Then compare your table with Table 1.4.

Table 1.4: The Population of Upper Volta, 1920–1960

Year	Population
1920	1,206,198
1930	1,372,061
1940	1,338,926
1950	1,404,228
1960	1,536,490

Ready to do another one? All right, here it comes. In the first quarter of 1999, Brooklyn International Corporation had sales of $45,000 in the United States and $38,000 abroad. In the second quarter sales in the United States were $58,000 and sales abroad were $61,000. Sales in the third quarter in the United States were $48,000 and sales abroad were $55,000. And in the fourth quarter sales in the United States were $62,000 and sales abroad were $73,000. Compile a table showing Brooklyn International's quarterly sales in the United States and abroad in 1999. Do this on a separate piece of paper and then check your work with mine in Table 1.5.

Table 1.5: Quarterly Foreign and Domestic Sales of Brooklyn International Corporation, 1999

Quarter	U.S.	Foreign	Total
1	$ 45,000	$ 38,000	$ 83,000
2	58,000	61,000	119,000
3	48,000	55,000	103,000
4	62,000	73,000	135,000
Total	213,000	227,000	440,000

Line Graphs

Most graphs are actually very simple, and line graphs are perhaps the easiest to draw. All you do is lay out the graph, place the dots, and then connect them. But we're getting a bit ahead of ourselves. Remember Table 1.1, which listed Lisa's weight? I'm going to show you how to take the data from that table and make it into a line graph. Let's start with Figure 1.1. It's a simple piece of graph paper on which I've drawn two axes. On the vertical axis we have weight, and on the horizontal axis we have age.

Figure 1.1: Graph with Two Axes

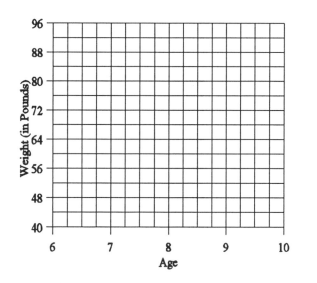

Now we put in the dots representing Lisa's weight at each age. We did that in Figure 1.2.

Figure 1.2: Graph with Dots Representing Lisa's Weight

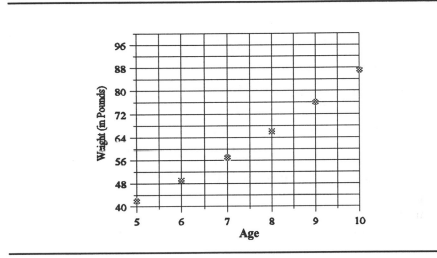

And finally we connect the dots. That gives us the line graph in Figure 1.3. Now exactly what does this graph show us?

Figure 1.3: Lisa's Weight from Age Five to Age Ten

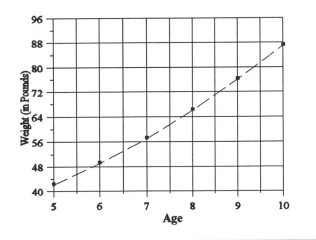

It shows us that Lisa's weight went up each year from when she was five until she was ten. We might even go so far as to say that her weight went up at an increasing rate each year. She gained 7 pounds, then 8 pounds, 9 pounds, 10 pounds, and 11 pounds. You can see all that just from looking at a graph. This may be just what the Chinese had in mind when they made up the proverb, "A picture is worth a thousand words." Indeed the word "graph" *means* "picture." So when you're asked to draw a graph, you're just being asked to draw a picture. And when someone looks at your graph, you hope they *get* the picture.

Before we do the next graph, I'd like you to notice that we began our vertical axis at a weight of 40 pounds. Would it have been wrong to start it at zero? No, but if we had, we would have had all that empty space between 0 and 40 that we weren't using. The important part of our graph—from 40 pounds up to 88 pounds—would have been squeezed into the upper half of the graph paper. Imagine that we were graphing weights from 648 pounds to 688 pounds. Would you *still* want to start at zero? If you did, your graph would be squeezed into a tiny space at the top of the graph paper. It would look like the graph in Figure 1.4.

Figure 1.4: Graphing Weights from 648 to 688 Pounds

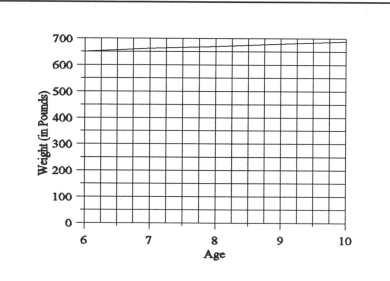

Here's another graph for you to do—the weights of Lisa and Karen. Use the data in Table 1.2.

Did you draw the graph? It should look something like what I've done in Figure 1.5.

Figure 1.5: Lisa's Weight and Karen's Weight from Age Five to Age Ten

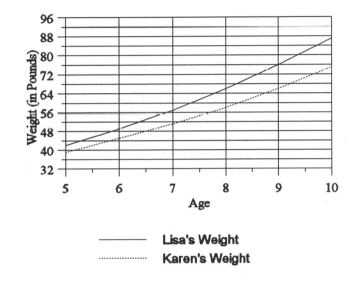

You'll notice that I used a dotted line to trace Karen's weight. That way there's no confusing the two girls' weights. Using a different color ink would be even better.

Bar Graphs

Using the same data that we used to make line graphs, we can also make bar graphs. In Figure 1.6 I've done a bar graph of Lisa's weight.

Figure 1.6: Lisa's Weight from Age Five to Age Ten

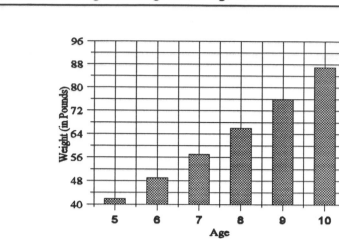

A bar graph can also be drawn horizontally. Using the same data, I've done that in Figure 1.7.

Figure 1.7: Lisa's Weight from Age Five to Age Ten

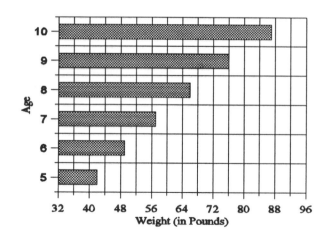

When do you draw a bar graph and when do you draw a line graph? Sometimes one or the other looks more dramatic. Suppose you were making a presentation at a sales conference. You wanted to show that sales were going through the roof, and you wanted to make sure that even the people way in the back of the room could see your flip chart. I'd go for the line graph, and I'd make that line with a thick red marker.

Enough chitchat! It's time to go back to work. Remember those Republican and Democratic representatives we left back in Table 1.3? Let's see you do a vertical bar graph, region by region. Use your own graph paper and lay out the regions on the horizontal axis and the number of seats won by Democrats and Republicans on the vertical axis.

Did you do it? How did it come out? I hope something like what I've got in Figure 1.8.

Figure 1.8: House Seats Won by Democrats and Republicans in the Election of 2020, by Region

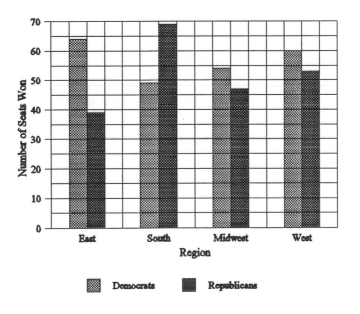

You'll notice a couple of things about my graph. First, I distinguished between the Democratic and Republican bars by making the Democrat bars a lighter shade. Once you know that the Democrats have the lighter–shaded bars and the Republicans have the darker–shaded bars, which I noted at the bottom of the graph, you can see the election results at a glance. Incidentally, you know you've drawn your graph well when the reader can see exactly what you're saying at just a glance.

You may have also noticed that I started the vertical axis at zero, rather than at around 35. Why? Because I wanted to compare the total number of seats won by the Republicans and Democrats in each region. If I started my vertical axis at 35, my graph would have had a truncated effect, exaggerating the relative regional strengths of the Democrats and Republicans. I've illustrated that in Figure 1.9.

Figure 1.9: House Seats Won by Democrats and Republicans in the Election of 2020, by Region

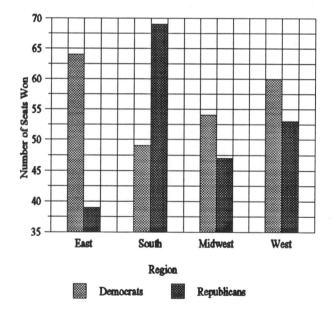

If you just glance at Figure 1.9, you get the impression that the Democrats completely dominate the East, Midwest, and West, while the Republicans completely dominate the South. But if you were to glance back at Figure 1.8, you would quickly see that the two parties are a lot more competitive in each region.

Does that mean that Figure 1.8 is more correct than Figure 1.9? No! A careful reader would get exactly the same message from both graphs, since both show exactly the same data. The problem is that most readers will not spend more than a few seconds looking at your graphs. Sorry, but that's how the world works.

To come back to the flip charts, you have only a few seconds to capture your audience's attention, so you need to make the most of it. You want to draw large, colorful charts that are easy to read from a distance. Whether you draw line graphs or bar graphs or present your data in other ways often comes down to judgment. And judgment comes with experience. That's why I want you to keep drawing graphs.

Histograms

Histograms are bar graphs that are pushed together. Glance back at Figure 1.6, which is a bar graph of Lisa's weight from ages five to ten. Using this data, we drew a histogram in Figure 1.10.

Figure 1.10: Lisa's Weight at Ages Five to Ten

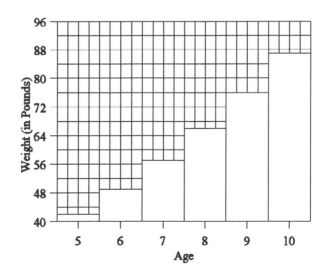

A histogram can be even more dramatic when it is done in a solid color, as in Figure 1.11. Remember that readers seldom spend more than a few seconds looking at graphs and charts, so you need to make a strong visual impression.

Figure 1.11: Lisa's Weight at Ages Five to Ten

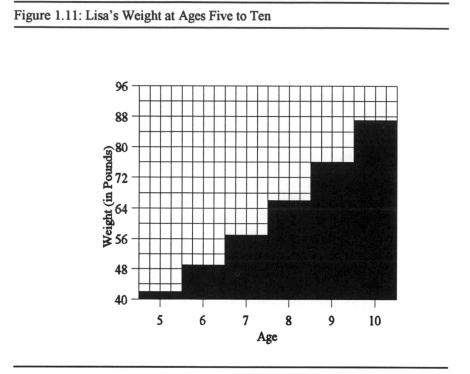

Now it's your turn to draw a histogram. Use the data in Table 1.4, which shows the population of Upper Volta from 1920 to 1960.

Figure 1.12: Population of Upper Volta, 1920–1960

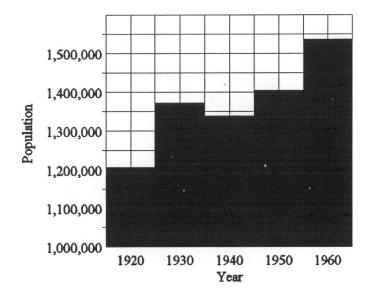

Your histogram should look something like mine in Figure 1.12. Mine is a bit more stretched out horizontally than the one I drew in Figure 1.11. There are no set rules about proportions, so you can stretch out a histogram vertically or horizontally. In a box near the end of the chapter, "How to Lie with Statistics," we talk more about this issue.

Pie Charts

Pie charts are exactly what they sound like—simply sliced–up pies. Each slice represents a portion of the pie. Usually we draw pie charts with three to six slices, although occasionally you may see one with eight or nine slices. The problem with having a lot of slices is that some may be too small to see very well.

The pie chart shown in Figure 1.13 has four equal slices, which may be called quadrants, or quarters.

Figure 1.13: Pie Chart with Four Equal Quadrants

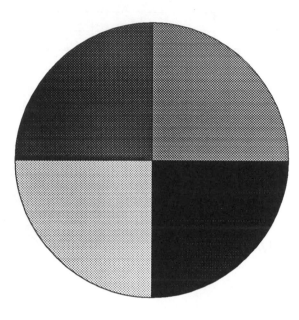

What would a pizza pie look like? It would have eight equal slices, unless we're talking about one of those square Sicilian pies. (I always thought pies were round and cakes were square.)

Data for pie charts need to be in percentage form—a topic we'll take up in Chapter 4. For now, we'll just put our data into percentage form. Suppose the federal government gets 38 percent of its revenue from the personal income tax, 34 percent from social insurance taxes, 15 percent from the corporate income tax, 9 percent from borrowing, and 4 percent from excise taxes. Draw a pie chart for these sources of federal revenue. (Do the best you can sketching a circle on a piece of paper or, better yet, use a compass.) Your pie chart should look something like the one I've drawn in Figure 1.14.

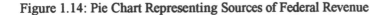

Figure 1.14: Pie Chart Representing Sources of Federal Revenue

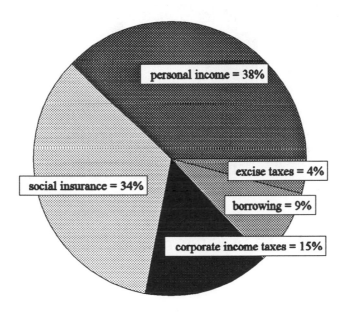

Complex Graphs

Don't worry. The graphs you'll be drawing won't be *that* complex. What we'll be doing here is making comparisons rather than just drawing individual pictures.

For example, let's compare the Republican, Democratic, and Independent votes in the presidential elections in 1992 with those in 1996. I've done that in the bar graph in Figure 1.15.

Figure 1.15: Percentage Share of 1992 and 1996 Presidential Votes

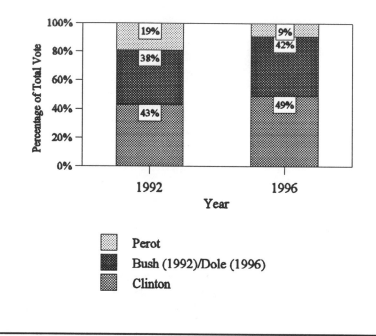

Use that data to draw two pie charts (one for 1992 and the other for 1996) that make the same comparison.

Your pie charts should look something like mine in Figure 1.16. Incidentally, which type of graph makes a clearer comparison—the bar graph in Figure 1.15 or the pie chart in Figure 1.16?

Figure 1.16: Percentage Share of 1992 and 1996 Presidential Votes

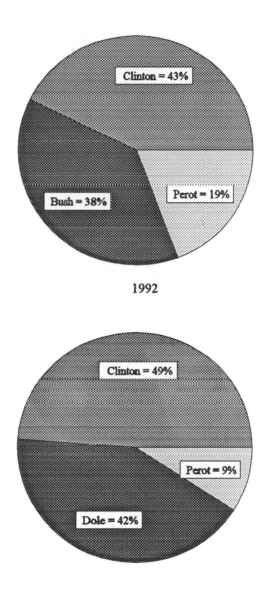

1992

1996

My own feeling is that the pie charts do a better job. As you draw more and more graphs, you can mentally make visual comparisons. You'll find this very useful, except, perhaps, when you're driving.

Now for our last trick. And to show you what a nice guy I am, I'll do this one by myself and I won't ask you to do another one like it.

Before we can even think about drawing this graph, I'd like you to take a look at Table 1.6. Do you see what's going on here?

Table 1.6: Federal, State, and Local Government Spending, 2025–2029 (in billions of dollars)

	2025	2026	2027	2028	2029
Federal	3,810	3,990	4,110	4,240	4,380
State	1,140	1,180	1,220	1,270	1,330
Local	1,680	1,750	1,830	1,920	2,020
Total	6,630	6,920	7,160	7,430	7,730

We've got a summary of total federal, state, and local government spending from 2025 through 2029. I know that governments like to plan ahead, but surely not *this* far ahead. If these numbers in thousands of billions are unfamiliar to you, you'll want to go over the box "Millions, Billions, and Trillions" below.

Millions, Billions, and Trillions

Our nation's output is measured in trillions of dollars. Our planet is populated by billions of people. And during your lifetime, you may earn millions of dollars. Can you write out, with all the necessary zeros, the numbers one hundred million, one hundred billion, and one hundred trillion? Do it right here:

One hundred million looks like this: 100,000,000. We write 100 followed by two sets of three zeros, separated by commas.

One hundred billion looks like this: 100,000,000,000. We write 100 followed by three sets of three zeros, separated by two commas.

And here's 100 trillion: 100,000,000,000,000. It's 100 followed by four sets of three zeros, separated by three commas.

Do you need to *know* all this? You do if you work a lot with very large numbers. Otherwise, whenever you come across these numbers, you can always refer back to this box.

What would a line graph of the data in Table 1. 6 look like? Picture a line graph with federal spending at the bottom, topped by state spending, and then local spending. Their sums would be the top line of the graph. That graph appears in Figure 1.17.

Figure 1.17: Federal, State, and Local Government Spending, 2025–2029

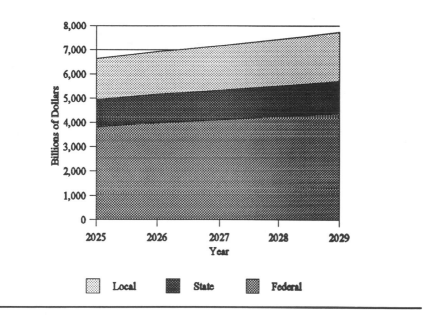

Notice that we have the three forms of government spending lined up vertically, with federal on the bottom, state in the middle, and local on top. The three add up to total government spending. Is this graph accurate? Check it against the data in Table 1.6.

You probably will never be called upon to draw a graph like this one, but you may well have to read graphs very much like it. What four conclusions can you read by studying the graph?

You can easily see that federal spending is rising, and with somewhat more difficulty you may observe that state and local spending are also going up. And finally, you can easily see that total government spending is rising.

Before we finish this chapter, we have one more thing to consider—the box "How to Lie with Statistics." We can use the same data to draw graphs that will lead observers to very different conclusions. Remember that most people just glance at graphs for a few seconds and can easily be led to the wrong conclusions. I will leave it up to you to decide whether this may involve lying.

How to Lie with Statistics*

This is actually the name of a best-selling book that shows you how you can trick people into believing that a given set of statistics looks good, bad, or indifferent. One of the best ways is by using the "gee-whiz graph." Using exactly the same set of data, the book shows these three graphs reproduced here as Figures 1.18, 1.19, and 1.20.

First compare Figures 1.18 and 1.19. In Figure 1.18 we begin the vertical axis at zero and in Figure 1.19 we begin it at 18. You'll notice that we seem to have a more pronounced rise in the line graph in Figure 1.19.

---

*Darrell Huff, *How to Lie with Statistics* (New York: Norton, 1993), Chapter 5.

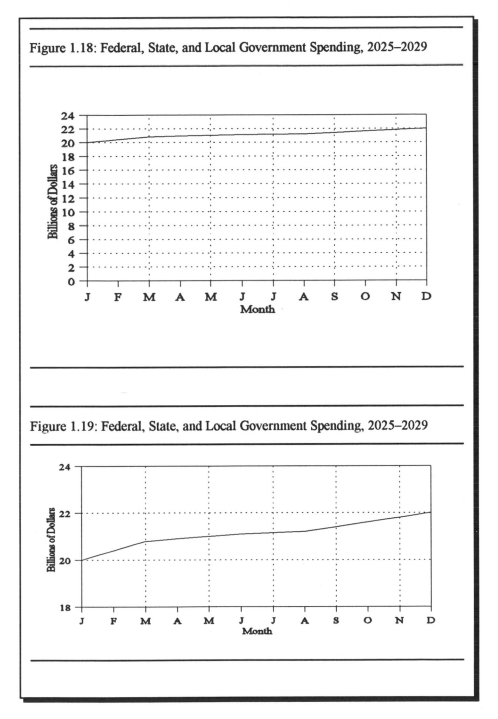

Figure 1.18: Federal, State, and Local Government Spending, 2025–2029

Figure 1.19: Federal, State, and Local Government Spending, 2025–2029

Figure 1.20: Federal, State, and Local Government Spending, 2025–2029

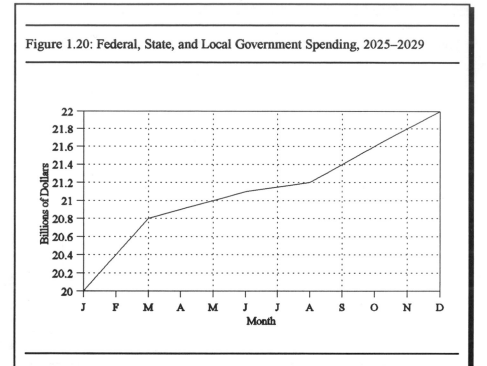

But you ain't seen nothin' yet! Take a look at Figure 1.20. Any casual observer could only be impressed with the rapid increase shown here. And yet, all we did was widen the vertical scale. This is truly a "gee-whiz" graph.

As you gain experience observing graphs, you will learn to recognize how easy it is to lie with statistics, especially if you play around with the vertical scale. And as you draw more of your own graphs, you may also grow more proficient at lying with statistics.

Chapter 2

Basic Arithmetic Operations

At the very beginning of the book I said you could do statistics as long as you could add, subtract, multiply, and divide. I lied. You will also need to convert fractions into decimals, round, multiply decimals, and do interpolation and extrapolation.

I could argue that these arithmetic processes are merely applications of addition, subtraction, multiplication, and division. But that's small consolation if you don't already know how to do them, which is exactly why I wrote this chapter. Of course, you may remember how to do all this stuff, in which case you may skip the entire chapter. Or your memory may need a little refreshing in at least some of these skills. Now, I'm not going to be looking over your shoulder, so you'll have to let your conscience be your guide. Just keep in mind that you'll need to use each of these arithmetic operations in later chapters.

Converting Proper Fractions into Decimals

We're going to be converting fractions like 3/8 into decimals. If you remember how to do this, then go directly to the next section.

To convert a proper fraction into a decimal we divide the bottom number (denominator) into the top number (numerator). Go ahead and convert 3/8 into a decimal.

Solution:
$$8\overline{)3.0^60^40}\quad\overset{.375}{}$$

Here's another one. Convert 4/5 into a decimal.

Solution:
$$5\overline{)4.0}\quad\overset{.8}{}$$

Do you always have to divide a denominator into the numerator to convert a fraction into a decimal? The answer is yes. But sometimes you can take a shortcut. We'll take that shortcut when we convert 7/10 into a decimal.

When we divide a number by 10, we move the decimal point one place to the left. For instance, we get .3 when we divide 3 by 10. What we really did was this:

$$\underset{\leftarrow}{3}.0$$

The fraction 7/10 may be read as 7 divided by 10. If we took the number 7, or 7.0, and divided it by 10, we'd end up with .7, or .70.

Now change the fraction 9/10 into a decimal. Did you get .9? Good. We're ready to move on to one-hundredths. The fraction 33/100 is read as thirty-three one-hundredths. Can you change it into a decimal?

Solution: $\dfrac{33}{100} = .33$

What we did was take the 33, or 33.0, and move the decimal two places to the left:

$$\underset{\leftarrow}{.33}.0$$

So when you want to divide a number by 100, just move the decimal two places to the left.

Let's see if you can convert each of these fractions into decimals:

(1) $\dfrac{1}{10}$ (2) $\dfrac{3}{10}$ (3) $\dfrac{7}{100}$ (4) $\dfrac{83}{100}$

Solutions:

(1)   .1   (2)   .3   (3)   .07   (4)   .83

So far we've done tenths and one-hundredths. One tenth is ten times the size of one one-hundredth. And one one-hundredth is ten times the size of one one-thousandth, which may be written as a fraction: 1/1000.

Can you convert the fraction 247/1000 into a decimal?

Solution: $\dfrac{274}{1000} = .247$

Now change the fraction 819/1000 into a decimal.

Solution: $\dfrac{819}{1000} = .819$

Here's one more set of problems for you to work out. Please convert each of these fractions into decimals:

(1) $\dfrac{6}{10}$ (2) $\dfrac{37}{100}$ (3) $\dfrac{693}{1000}$ (4) $\dfrac{18}{100}$

(5) $\dfrac{349}{1000}$ (6) $\dfrac{24}{1000}$ (7) $\dfrac{2}{100}$ (8) $\dfrac{3}{1000}$

Solutions:

(1)   .6   (2)   .37   (3)   .693   (4)   .18

(5) .349   (6)   .024   (7)   .02   (8)   .003

Rounding

If you earned $46,127.32, how much would your earnings be rounded to the nearest thousand dollars? It would be $46,000. And how much would your earnings be rounded to the nearest hundred dollars?

It would be $46,100. And how much would your earnings be rounded to the nearest dollar?

It would be $46,127. We do a lot of rounding in statistics, so we need to be clear about how to round off our numbers. Most of the rounding we'll do will be with decimals. Round off this decimal to the nearest tenth: .1649.

The answer is .2. And how much would .1649 be rounded to the nearest hundredth?

It would be .16. And finally, how much would .1649 be rounded to the nearest thousandth?

The answer is .165. Now how would you round off .05 to the nearest tenth?

Usually we would round it off to .1. We would call that rounding up. Fives, however, can be rounded up or down. Some statisticians will alternate rounding their fives up and down. For example, suppose they were rounding each of these decimals to the nearest hundredth: .005; .135; .955; .615.

Their answers would be: .01; .13; .96; and .61. However, most people round all their fives up. So they would have rounded these decimals *this* way: .01; .14; .96; and .62. Which way is right? They both are.

If you can round off each of these decimals, then you definitely have rounding down.

Problems:

(1)   Round off each of these decimals to the nearest whole number:

    (a)   1.49   (b)   0.6   (c)   0.47   (d)   4.649

(2)   Round off each of these decimals to the nearest tenth:

     (a)   .001   (b)   1.349   (c)   0.662   (d)   0.954

(3)   Round off each of these decimals to the nearest hundredth:

     (a)   .04269   (b)   1.9665   (c)   .1458   (d)   .96457

(4)   Round off each of these decimals to the nearest thousandth:

     (a)   10.13445 (b)   .91864   (c)   1.80257   (d)   .75098

Solutions:

(1)	(a)	1	(b)	1	(c)	0	(d)	5
(2)	(a)	0	(b)	1.3	(c)	0.7	(d)	1.0
(3)	(a)	.04	(b)	1.97	(c)	.15	(d)	.96
(4)	(a)	10.134	(b)	.919	(c)	1.803	(d)	.751

Multiplying Decimals

We do a lot of multiplying of decimals in statistics, especially when we deal with percentiles (Chapters 6 and 8). Unfortunately, multiplying with decimals probably leads to more mistakes than any other arithmetic operation. But if you follow just one simple rule, you can avoid all that grief. We'll work out one problem and then apply the rule:

Problem:

$$2.14 \times 3.5$$

Solution:

$$\begin{array}{r} 2.14 \\ \times 3.5 \\ \hline 1070 \\ 642 \\ \hline 7.490 \end{array}$$

This was really a two-part problem. The first part was straight multiplication. The second part was to figure out where to place the decimal point.

Here's the rule: Add the number of places to the *right* of the decimal in the numbers being multiplied and place the decimal point that number of places to the *left* in the answer. For instance, 2.14 has two numbers to the right of the decimal point, and 3.5 has one number to the right of the decimal point. That gives us three numbers to the right of the decimal point. In our answer, we placed our decimal point three places to the left. After you have a few problems under your belt, you'll be able to place your decimal points without even thinking about it.

Problem: See what you can do with this problem:

$$14.06 \times 8.25$$

Solution:

```
  14.06
 ×8.25
  7030
 2812
11248
115.9950
```

If you'd like some more practice, here's a whole bunch of problems for you to work out:

(1) 3.5   (2) 10.17 (3) 31.76 (4) 1.09
    ×6.2   ×9.84   ×59.81   ×7.53

(5) 1.183 (6) .301   (7) 15.926 (8) 5.1164
    ×.72   ×.952   ×10.743   ×2.065

Solutions:

(1) 3.5   (2) 10.17   (3) 31.76   (4) 1.09
    ×6.2   ×9.84   ×59.81   ×7.53
     70   4068   3176   327
    210   8136   25408   545
    21.70   9153   28584   763
              100.0728   15880   8.2077
                             1,899.5656

(5) 1.183 (6) .301   (7) 15.926   (8) 5.1164
    ×.72   ×.952   ×10.743   ×2.065
    2366   602   47778   255820
    8281   1505   63704   306984
    .85176   2709   111482   1023280
               .286552   159260   10.5653660
                             171.093018

Interpolation

Interpolation is used to find an unknown number that is situated between two known numbers. Suppose you needed to find Y in Table 2.1 when X is 2.15. See if you can do it.

Table 2.1: Finding Y

X	Y
2.0	31
2.1	38
2.2	46
2.3	56
2.4	70
2.5	89

Since 2.15 is exactly halfway between 2.0 and 2.1, we want to find a Y that is exactly halfway between 38 and 46. So Y = 42.

Now suppose X were 2.24. How much would Y be?

Solution: 2.24 is 4/10, or .4, of the distance between 2.2 and 2.3, so we want to find a Y that is .4 the distance between 2.2 and 2.3, so we want to find a Y that is .4 of the distance between 46 and 56. The distance between 46 and 56 is 10, so .4 × 10 = 4. 46 + 4 = 50.

Here's one more. If X = 2.47, how much is Y?

Solution: 2.47 is 7/10, or .7, of the distance between 2.4 and 2.5. The distance between the Y's is 19. .7 × 19 = 13.3. 13.3 + 70 = 83.3.

We will be doing a lot of interpolating in Chapter 6 when we work with percentiles. See if you can do *this* percentile problem now even though we'll hold off on defining percentiles until Chapter 6. An IQ of 120 is in the 30th percentile and an IQ of 130 is in the 45th percentile. In what percentile is an IQ of 123?

Solution: An IQ of 123 is 3/10 of the way between an IQ of 120 and 130. So we're looking for a percentile that is 3/10 of the way between the 30th and the 45th. .3 × 15 = 4.5. 4.5 + 30 = 34.5, or the 34.5 percentile.

Let's do one more. A weight of 100 pounds is at the 50th percentile and a weight of 120 is at the 85th percentile. At what percentile is a weight of 118?

Solution: A weight of 118 is 18/20, or 9/10, of the distance between 100 pounds and 120 pounds. .9 × 35 = 31.5. 31.5 + 50 = 81.5 percentile.

Extrapolation

   Whereas interpolation involves finding a number situated between two known numbers, extrapolation involves finding a number, or numbers, beyond a series of known numbers. Once you've discerned a trend, you project what comes next.

   As the Great Depression wore on through the 1930s, most Americans thought the bad economic times would continue indefinitely. And during the high inflation years of the late 1970s and early 1980s, most people believed prices would keep rising by 8 or 10 percent a year. When we see a trend, we tend to extrapolate by projecting that trend into future years.

   Let's start with a very simple extrapolation. Fill in the next three numbers of this series: 2, 4, 6, 8, __, __, __.

   I know you wrote 10, 12, 14. Here's one that's somewhat harder: 1, 2, 4, 7, 11, __, __, __. Please fill in the blanks.

   The next three numbers would be 16, 22, and 29. Here's the trend: 1 (+1), 2 (+2), 4 (+3), 7 (+4), 11 (+5), 16 (+6), 22 (+7), 29.

   Sometimes we can extrapolate a trend that we detect on a graph. See if you can do that with the line graph drawn in Figure 2.1 and then compare your results with mine in Figure 2.2.

Figure 2.1: Graph for Extrapolating Trend

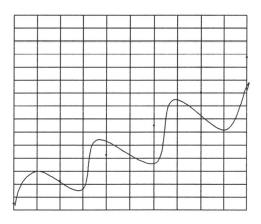

Figure 2.2: Graph

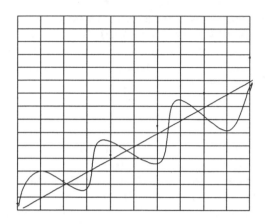

This type of analysis will be used extensively in Chapter 10 (Correlation) and Chapter 11 (Prediction and Regression). We do need to remember, however, that we cannot always assume that present trends will continue indefinitely. If they did, we would all have made millions in the stock market.

Fast Multiplication and Division

Fast Multiplication

How much is 43 × 10? It's 430. All we did was tack on a zero to 43. Here are a few more problems:

Problems: Multiply these numbers by 10:

(1)   10   (2)   136   (3)   91   (4)   3,065

Solutions:

(1)   100   (2)   1,360   (3)   910   (4)   30,650

You probably noticed that when we multiplied by 10, what we were really doing was adding a zero to the number being multiplied. That observation is fine as long as we're dealing with whole numbers. But when we deal with decimals, we've got to worry about the decimal point.

How much is .8 × 10? The answer is 8. How did we get that? We moved the decimal point one place to the right. But what if we had merely added a zero? Then we would have gotten .80, which is equal to .8. So when you're multiplying a decimal by 10, you need to make sure to move the decimal point one place to the right.

Problems: Multiply these numbers by 10:

(1)   1.4   (2)   .5   (3)   .03   (4)   10.16

   Solutions:

(1)   14   (2)   5   (3)   .3   (4)   101.6

For our next trick we're going to multiply by 100. To do that we add two zeroes to whole numbers and move the decimal point two places to the right if we have decimals.

Problems: Multiply each of these numbers by 100:

(1)   17   (2)   208   (3)   1,294

(4)   .01   (5)   .746   (6)   3.09

   Solutions:

(1)   1,700   (2)   20,800   (3)   129,400

(4)   1 (or 1.0)   (5)   74.6   (6)   309 (or 309.0)

If we were to multiply a number by 1,000, we'd need to add three zeros if it were a whole number. And if it were a decimal, we'd move the decimal three places to the right. As Ross Perot has often said, it's as simple as that. And it's *that* simple to divide by tens, hundreds, and thousands.

Fast Division

When we divide by tens, hundreds, and thousands, we do exactly the opposite of what we did when we multiplied. See what you can do with this problem set:

Problems: Divide each of these numbers by 10:

(1)   700   (2)   16   (3)   38.6   (4)   .05

   Solutions:

(1)   70   (2)   1.6   (3)   3.86   (4)   .005

Now we'll be dividing by 100. All you need to do is move the decimal point two places to the left or, in the case of whole numbers ending with at least two zeros, lop off the last two zeros.

Problems: Divide each of these numbers by 100:

(1)   18.2   (2)   56   (3)   5,000   (4)   7

(5)   .09   (6)   1.6   (7)   .008   (8)   200

       Solutions:

(1)   .182   (2)   .56   (3)   50   (4)   .07

(5)   .0009   (6)   .016   (7)   .00008   (8)   2

To divide by 1,000, we need to move the decimal point *three* places to the left. Do you ever need a calculator do multiply or divide by 10, 100, or 1,000? Not only do you *not* need a calculator to carry out these operations, but using a calculator to do this type of multiplication and division can lead to errors. It's easy to make mistakes punching numbers with decimals or a bunch of zeros into a calculator.

Manipulating Negative Numbers

If you visited the casinos at Atlantic City, New Jersey, for four straight weekends and lost $150 on the first weekend, won $400 on the second weekend, lost $270 on the third, and won $90 on the fourth, how much are your total winnings over the four weekends?

Solution: $-\$150 + \$400 - \$270 + \$90 = +\$70.$

Now we'll do a countdown. I'll supply the first three numbers and you supply the next three:

$$4, 3, 2 \_\_, \_\_, \_\_\_.$$

The full countdown would be 4, 3, 2, 1, 0, -1. You'll notice that negative numbers are denoted by minus signs, but we don't usually bother to place + signs in front of positive numbers. The only time we do is when there are some negative numbers close by.

Now add these numbers: -2, -6, and -3.

The answer is -11. Now add -4, -9, $+6$, -1, and $+4$.

The answer is -4. Add the positive numbers ($+6$ and $+4 = +10$). Then add the negative numbers (-4 and -9 and $-1 = -14$.) Then add the positive numbers and the negative numbers. $+10 + (-14) = +10 - 14 = -4$.

How much is $+6$ take away $+8$?

The answer is -2. $6 - 8 = -2$. How much is $+5$ take away -9?

The answer is $+4$. $5 - (-9) = 5 + 9 = 14$.

Why are we manipulating negative numbers? Because in the next chapter we'll need that skill finding means, medians, modes, and ranges.

Chapter 3

The Mean, Median, Mode, and Range

The whole point of statistics is to analyze data. Suppose we knew the annual incomes of 25 people. We could find their average, or mean, income. If we arranged their incomes in an array, from the lowest to the highest, we could find their median, or middle, income, which would be the 13th income in that array. We could find the mode, or the income that is mentioned more than any other income. And finally, we could find the range by subtracting the lowest income from the highest.

We'll be working on mean, median, mode, and range problems separately, and then, for our grand finale, we'll work out problem solving for all of them using the same data. I think you'll find this a very easy chapter, and you may even use a calculator.

The Mean

Simple Mean

You might not realize it, but you've calculated the mean many times. If you received exam grades of 90, 65, and 80, calculate your mean to the nearest tenth:

Solution: $90 + 65 + 80 = 235$

$$\frac{235}{3} = 78.3$$

So to find the mean, just add up the values and divide by the number of values. You probably don't need a formula to solve these problems, but we have one anyway, and we'll need it when we get to standard deviation in Chapter 7. That formula is:

$$\overline{X} = \frac{\Sigma X}{n}$$

$\overline{X}$ is the mean of X. Σ is the Greek letter sigma, and means "the sum of." In this case we're talking about the sum of the X's (which would be 235). And n stands for the number of terms (which would be 3).

Do you need to write this formula every time you're finding the mean? No, but you will need to remember what each of the terms is and be ready to use the formula when you get to Chapter 7.

Problem: Find the average IQ to one decimal place for seven people who have IQs of 87, 94, 99, 102, 115, 121, and 134.

Solution: $\overline{X} = \dfrac{\Sigma X}{n} = \dfrac{752}{7} = 107.4$

Problem: Use the data in Table 3.1 to find the mean unemployment rate (to the nearest tenth) for the year 2008.

Table 3.1: Monthly Unemployment Rates for 2008

Month	Rate
January	5.9
February	6.1
March	6.4
April	6.6
May	6.5
June	6.7
July	6.6
August	6.3
September	6.3
October	6.2
November	6.0
December	6.1

Solution: $\overline{X} = \dfrac{\Sigma X}{n} = \dfrac{75.7}{12} = 6.3$

Problem: Use the data in Table 3.2 to find Borders' average monthly sales (to the nearest penny).

Table 3.2: Monthly Sales of the Belton, Texas, Borders Bookstore, 2015

Month	Rate
January	$1,972,356.12
February	1,416,903.43
March	2,006,431.09
April	2,013,463.98
May	2,134,802.00
June	2,397,164.48
July	2,247,935.82
August	2,190,838.17
September	2,462,190.29
October	2,530,145.60
November	2,096,873.55
December	2,889,403.74

Solution: $\overline{X} = \dfrac{\Sigma X}{n} = \dfrac{\$26,358,508.27}{12} = \$2,196,542.36$

I think you've got enough problems under your belt to do a problem set.

Problems: For each of these, find the mean to one one-hundredth.

(a)   Here are the weight losses of the class of March, 2017 at the Bangor Quick Weight Loss Club: –8, –4, –10, +2, –7, 0, –5, –10, +3, –12.

(b)   Winnings and losses of the Optimists' Bingo Club: –$12, +$40, +$17, —$31, +$127, –$9, –$15, +$22, +$86, –$38.

(c)   Use the data in Table 3.3:

Table 3.3: Temperature at 6 A.M. in St. Paul, MN, in February 2022 (in degrees Fahrenheit)

Day	Temperature	Day	Temperature
1	−13	15	−8
2	−8	16	−7
3	−15	17	−2
4	−2	18	8
5	6	19	5
6	0	20	0
7	−7	21	6
8	−9	22	−4
9	−13	23	−10
10	−7	24	−17
11	3	25	−18
12	0	26	−12
13	6	27	−6
14	−1	28	−1

(d)   Use the data in Table 3.4:

Table 3.4: Qualifying Times for Quarter Mile College Women's Finals, May 24, 2037 (in seconds)

58.3	56.1	58.9	56.5
57.8	55.6	59.4	56.1
59.6	60.4	57.0	54.9
57.8	58.4	55.7	56.1
58.3	54.7	55.9	58.5
56.0	59.3	57.6	55.5
59.1	56.0	60.2	59.9
56.2	55.7	57.8	60.0
59.2			

Solutions:

(a) $\overline{X} = \dfrac{\Sigma X}{n} = \dfrac{-51}{10} = -5.10$ pounds

(b) $\overline{X} = \dfrac{\Sigma X}{n} = \dfrac{\$153}{10} = \$15.30$

(c) $\overline{X} = \dfrac{\Sigma X}{n} = \dfrac{-160 + 34}{28} = \dfrac{-122}{28} = -4.36$ degrees

(d) $\overline{X} = \dfrac{\Sigma X}{n} = \dfrac{1898.5}{33} = 57.53$ seconds

Weighted Average

Whether you realize it or not, you may already be familiar with weighted averages. In college, your grade point average, or GPA, happens to be a weighted average. Virtually every college computes it the same way: An A is a 4, a B is a 3, a C is a 2, a D is a 1, and an F is a 0. So if you got all A's, you'd have a perfect 4.0 GPA.

But you must also consider the number of credits you received in each course. Some courses are four credits, most happen to be three, some are two, one, or even one and a half credits or half a credit. Here's how you did last term:

Course	Credits	Grade
Intro to Accounting	4	D
Macroeconomics	3	C
Oral Communications	2	A
Intro to Psychology	3	B
Marketing	3	B
	15	

See if you can figure out your GPA. You'll need to multiply the credits by the grade, add up these numbers, and then divide by the total number of credits. After you've done that, compare your work with my solution.

Course	Credits	Grade			
Intro to Accounting	4	D	4×1	=	4
Macroeconomics	3	C	3×2	=	6
Oral Communications	2	A	2×4	=	8
Intro to Psychology	3	B	3×3	=	9
Marketing	3	B	3×3	=	9
	15				36

$\dfrac{36}{15} = 2.4$

Now please do this problem set:

Problems:

(a)   We want to find the weighted mean hourly wage of the Schwartz family. Denise Schwartz worked 46 hours, earning $18.40 an hour. David Schwartz worked 39 hours, earning $17.83 an hour. And Evan Schwartz worked 24 hours, earning $10.53 an hour.

(b)   The Shady Rest Pet Cemetery had 21 sales representatives who worked on straight commission. Two reps sold 2 burial plots each, 4 reps sold 3 plots each, 5 reps sold 4 plots each, 5 reps sold 5 plots each, 4 reps sold 6 plots each, and 1 rep sold 7 plots. What is the weighted average sales of the reps?

(c)   In Alpha Company, 8 recruits were 17 years old, 12 were 18 years old, 7 were 19 years old, 4 were 20 years old, 2 were 21 years old, and 1 was 22 years old. What is the weighted average of their ages?

Solutions:

(a)

	Hours		Hourly Earnings		Total Earnings
Denise	46	×	$18.40	=	$ 846.40
David	39	×	17.83	=	695.37
Evan	24	×	10.53	=	252.72
	109				$1,749.49

$$\frac{\$1,749.49}{109} = \$16.46$$

(b)

Number of Plots		Number of Reps		Total Sales
2	×	2	=	4
3	×	4	=	12
4	×	5	=	20
5	×	5	=	25
6	×	4	=	24
7	×	1	=	7
		21		92

$$\frac{92}{21} = 4.4 \text{ plots}$$

(c)

Age		Number of Recruits	Total Years of Age
17	×	8	136
18	×	12	216
19	×	7	133
20	×	4	80
21	×	2	42
22	×	1	22
		34	629

$\dfrac{629}{34}$ = 18.5 years old

The Median

I can still remember my first-grade teacher, Mrs. Anker, commanding the boys to stand in size places near the windows and the girls to stand in size places near the wardrobe. The boy with median height in that lineup was standing smack in the middle, as was the girl with median height. When a group of numbers is arranged in order of size, the median is the middle number. You can't miss it. For instance, how much is the median in this group of numbers?

2, 19, 22, 28, 61

The median is 22. What if the numbers are not in order of size? You would need to put them in size order and then find the median. For example, find the median in this group:

119, 4, 97, 101, 15, 80, 36

Solution:

4, 15, 36, 80, 97, 101, 119

The median is 80.

What if there is an even number of terms? See if you can find the median here, doing a bit of averaging:

9, 17, 24, 30, 57, 62

Solution:

The median would be exactly halfway between the third number, 24, and the fourth number, 30. So the median is 27.

Are you ready for another problem set?

Problems:

(a)   8, 14, 23, 28, 34, 42, 49, 57, 66

(b)   132, 18, 75, 39, 52, 99, 6

(c)   10, 24, 38, 43, 50, 77

(d)   97, 38, 44, 9, 103, 25

   Solutions:

(a)   34

(b)   6, 18, 39, 52, 75, 99, 132

         Median = 52

(c) $\dfrac{38+43}{2} = \dfrac{81}{2} = 40.5$

(d)   9, 25, 38, 44, 97, 103

         $\dfrac{38+44}{2} = \dfrac{82}{2} = 41$

Before we go any further, if you had to choose just one, would you use the median or the mean? For most distributions these two statistical tools have nearly identical numerical values. In the box below, we discuss cases in which the mean and the median are far apart.

The Mean versus The Median

   The mean, or average, may well be the most familiar and widely used statistical tool. Unfortunately, it sometimes provides a biased picture of the data. This usually happens when relatively few values pull up the mean. In the following distribution, just one extremely high value does this. Please calculate the mean for the people earning these incomes:

   $10,000, $10,000, $10,000, $10,000, $10,000, $10,000, $10,000, $10,000, $10,000, $1,000,000

   Solution: $\overline{X} = \dfrac{\Sigma X}{n} = \dfrac{\$1,090,000}{10} = \$109,000$

   The ten people earn an average income of $109,000, but nine of them earn just $10,000. This is another great example of how to lie with statistics.
   So what can we do? What we can do is use the median income instead of the mean income. The median is not subject to distortion by extreme values. How much is the median? It's $10,000.
   Here's another example. There's a beauty contest and all the finalists are

quite beautiful. Would any qualify as a perfect 10? Perhaps. Find the mean and median if the eight judges came up with these scores for one contestant:

9.9, 9.7, 9.8, 9.6, 9.7, 9.9, 9.7, and 0

$$\overline{X} = \frac{\Sigma X}{n} = \frac{68.3}{8} = 8.54$$

median = 9.75

Here we have an extremely *low* value that pushes the mean way down. If we can just get rid of the extreme values, then we can still use the mean. This is often done not just at beauty contests, but at athletic events like swim meets or ice skating events, where the high score and the low score are disregarded. We'll talk further about the relationship between the median and the mean when we get to Chapter 8, which deals with the normal distribution.

The Mode

The mode is the most frequent value in a set of numbers. What's the mode of this set?

3, 29, 19, 16, 90, 37, 3

The mode is 3 because that number occurs twice, and each of the others occurs only once.

What's the mode in this distribution?

10, 6, 15, 10, 30, 5, 6, 13, 12, 27, 10, 19

The mode is 10 because it occurs three times, while 6 occurs twice. Three wins over two every time.

Sometimes we have *bimodal* distributions, when two different numbers are both modes. Pick out the two modes here:

7, 3, 4, 7, 0, 9, 4, 2

Obviously 4 and 7 are modes in this bimodal distribution. Now let's see what you can do with this distribution:

31, 16, 4, 13, 22, 19, 7, 16, 34, 4, 21, 18, 2, 37, 24, 22, 15, 6, 11

It takes a bit of work, but the answers are 4, 16, and 22. We call this a *trimodal* distribution because there are three modes. Can there be *more* than three modes? Definitely. But such distributions are highly unusual.

Here's another problem set.

Problems: In each problem, find the mode or the modes:

(a)   −3, 9, 17, 0, 3, 10, 13, 9, 2

(b)   2, 11, −6, 15, 3, −2, 7, 18, 11, 4, −6, 16, 14, −5, 21, 4, −4, −6, 16

(c)   7, 29, 5, 18, 23, 14, 7, 12, 2, 0, 6, 13, 5, 1

(d)   8, 4, 17, 2, 19, 24, 10, 8, 7, 6, 25, 10, 15, 11, 27, 2

Solutions:

(a)   9

(b)   −6

(c)   5, 7

(d)   2, 8, 10

The Range

The range of a distribution of numbers is the difference between the highest and lowest values. For example, find the range of this array of numbers:

12, 19, 50, 62, 95, 116, 130, 157

The range is found this way: 157 − 12 = 145.

What if the numbers are not in order? Put them in order first, and then find the range. See what you can do with this group of numbers:

46, 16, 39, 171, 2, 98, 74, 46, 7, 100, 25

Solution: 171 − 2 = 169

Do you really need to put all the numbers in order to figure out the range? Not really. All you need to do is find the highest number and the lowest number, and then do the subtraction. Then why bother to put the numbers in order? Because we'll often want to find the mean, median, and mode, as well as the range, and putting the values in numbered order makes those jobs a lot easier. And we'll need to do jobs like that in the following section. But right now, I'd like you to do this problem set:

Problems: Please find the range for each of these numerical distributions:

(a)   15, 96, 37, 55, 29, 3, 119, 84, 61, 3, 71, 104, 90, 110

(b)   40, 0, 15, −12, 78, 32, 88, 145, 13, −18, 72, 159, 99, 6

(c)   46, 70, 11, −54, 9, −39, 2, 83, 54, 96, −15, 12, 76, 58, 40, −38, 65

(d)   −13, −9, 4, 19, −30, 82, 74, 107, −24, 0, 29, 43, −48, 67, 93, 118

Solutions:

(a)   119 − 3 = 116

(b)   159 − (−18) = 159 + 18 = 177

(c)   96 − (−54) = 96 + 54 = 150

(d) $118 - (-48) = 118 + 48 = 166$

Finding the Mean, Median, Mode, and Range

Let's put it all together now. First I'll give you an array, which is a distribution of numbers that are listed in order of magnitude, and then I'll give you numerical distributions that need to be put into arrays before we can find the mean, median, and mode. (As we mentioned in the last section, you can easily find the range without forming an array.)

Problem: Find the mean (to two decimal places), median, mode, and range of this array:

$$68, 14, 29, 53, 72, 5, 14, 86, 12$$

Solution:

$$5, 12, 14, 14, 29, 53, 68, 72, 86$$

$$\overline{X} = \frac{\Sigma X}{n} = \frac{353}{9} = 39.22$$

$$\text{median} = 29$$

$$\text{mode} = 14$$

$$\text{range} = 86 - 5 = 81$$

We'll close out this chapter with one last problem set:

Problems: For each of these numerical distributions, find the mean, median, mode, and range.

(a)   5, 24, −1, 17, −8, 2, 10, −1, 6, 15, −4

(b)   32, −7, 26, 10, −21, −4, 0, 16, −5, 12, −7, 3, 14

(c)   8, 46, −3, 19, 38, 7, −8, 12, −17, 19, −22, −4

(d)   Use the data in Table 3.5:

Table 3.5: Monthly Net Income of the Happy Day Car Repossession Company, 2027 and 2028 (in millions of dollars)

Month	2027	2028
January	$ −1.5	$ 0.3
February	4.2	15.7
March	7.7	6.2
April	−3.0	−1.5
May	10.1	2.5
June	6.3	−0.8
July	1.4	4.0
August	8.6	2.9
September	−1.9	13.4
October	−12.3	10.7
November	−7.2	4.4
December	1.6	1.3

Solutions:

(a) $\overline{X} = \dfrac{-14 + 79}{11} = \dfrac{65}{11} = 5.91$

−8, −4, 1, −1, 2, 5, 6, 10, 15, 17, 24
median = 5
mode = −1
range = 24 − (−8) = 24 + 8 = 32

(b) −21, −7, −7, −5, −4, 0, 3, 10, 12, 14, 16, 26, 32

$\overline{X} = \dfrac{113 - 44}{13} = \dfrac{69}{13} = 5.31$

median = 3
mode = −7
range = 32 − (−21) = 32 + 21 = 53

(c)   −22, −17, −8, −4, −3, 7, 8, 12, 19, 19, 38, 46

$$\overline{X} = \frac{149 - 54}{12} = \frac{95}{12} = 7.92$$

median $= \dfrac{7+8}{2} = \dfrac{15}{2} = 7.5$

mode $= 19$

range $= 46 - (-22) = 46 + 22 = 68$

(d) $\overline{X} = \dfrac{101.3 - 28.2}{24} = \dfrac{73.1}{24} = 3.06$

−12.3, −7.2, −3.0, −1.9, −1.5, −1.5, −0.8, 0.3, 1.3, 1.4, 1.6, 2.5, 2.9, 4.0, 4.2, 4.4, 6.2, 6.3, 7.7, 8.6, 10.1, 10.7, 13.4, 15.7

median $= \dfrac{2.5 + 2.9}{2} = \dfrac{5.4}{2} = 2.7$

mode $= -1.5$

range $= 15.7 - (-12.3) = 15.7 + 12.3 = 28.0$

Chapter 4

Frequency Distribution

When we collect data, whether from surveys, censuses, or casual observations, we end up with a mass of numbers, perhaps like the one shown in Table 4.1. Whatever tale these numbers tell cannot be read until they are reorganized into a more useful summary form. In this chapter we'll be reorganizing raw data into tabular and graphical frequency distributions.

Frequency Distributions

What do you do with a mass of unorganized data like the one shown in Table 4.1? You reorganize it into a grouped frequency distribution such as the one shown in Table 4.2. See if you can follow what I did here, because in a minute it will be your turn. You know how things work around here. I show you how to do a problem and then I ask you to do a similar problem.

Table 4.1: Hourly Wages Paid to Employees at the Friendly Financial Corporation, December 19, 2038

$5.90	7.35	10.64	7.73	5.96	8.90	9.32	6.50
8.25	11.65	8.45	5.95	8.03	6.32	12.84	10.00
7.55	9.47	8.85	12.50	14.16	9.95	7.05	8.16
9.84	11.75	5.90	6.75	8.99	6.05	8.83	10.50
13.87	9.90	5.95	8.75	6.43	7.82	6.50	13.00
17.10	9.08	5.97	6.89	6.45	9.75	13.40	6.80
8.20	9.57						

Table 4.2: Frequency Distribution of Hourly Wages Paid to Employees at the
Friendly Financial Corporation, December 19, 2038

Class Interval	Tally	Frequency
$5.00 – $5.99	JHT /	6
6.00 – 6.99	JHT ////	9
7.00 – 7.99	JHT	5
8.00 – 8.99	JHT JHT	10
9.00 – 9.99	JHT ///	8
10.00 and over	JHT JHT //	12

We took the hourly wages of employees at the Friendly Financial Corporation
and placed them in six groups, called class intervals. The highest group is
open–ended, since these people could well be earning $15 or $20 an hour, or even
more.

When we make up class intervals, we don't want them to overlap. They would
if we had intervals of $5 – $6, $6 – $7, $7 – $8, and so forth. This is something we
want to be very careful about. Actually, since I'll be making up all the class
intervals, this will be something *I'll* have to be very careful about, but when you're
working with frequency distributions out there in the real world, always watch out
for those overlaps.

By organizing our data we are able to get a much better picture of the wage
distribution at the Friendly Financial Corporation. Of the 50 hourly employees (we
find that total by adding the frequencies), 30 make at least $8 an hour, and 12 make
at least $10 an hour.

After you've had the chance to do your own frequency distribution, we'll be
drawing graphs from our frequency distribution tables. So let's see what you can
do with the data in Table 4.3 to complete Table 4.4. Then compare your work with
mine in Table 4.5.

Table 4.3: IQs of Freshman Class at the New England School for the Elite, September 18, 2041

119	154	133	121	128	139	116	140	127	135
112	143	124	159	132	117	141	150	137	162
126	118	149	142	120	164	123	109	135	151
110	119	145	134	122	164	149	108	135	140
146	120	113	129	136	128	147	130	122	150
118	143	120	134	148	153	120	115	140	136
130	123	146	138	119	124	140	126	141	137
116	147	135	161	133	157	129	126	116	121
155	138	123	114	128	132	156	149	120	113
140	127	122	136	114	152	145	126	130	151

Table 4.4: Frequency Distribution of IQs at the New England School for the Elite, September 18, 2041

Class Interval	Tally	Frequency
100–109		
110–119		
120–129		
130–139		
140–149		
150–159		
160–169		

Table 4.5: Frequency Distribution of IQs at the New England School for the Elite, September 18, 2041

Class Interval	Tally	Frequency
100–109	//	2
110–119	ЦНТ ЦНТ ЦНТ /	16
120–129	ЦНТ ЦНТ ЦНТ ЦНТ ЦНТ /	26
130–139	ЦНТ ЦНТ ЦНТ ЦНТ /	21
140–149	ЦНТ ЦНТ ЦНТ ЦНТ	20
150–159	ЦНТ ЦНТ /	11
160–169	////	4
		100

It looks as though a lot of smart kids go to that school. But, hey! It's an elite school that doesn't take in just anyone. At most other schools you'll find a much wider distribution of IQ scores, with most concentrated in the 90 to 110 range. Indeed, we'll spend all of Chapter 8 looking at the normal distribution, which is depicted graphically as the normal curve, sometimes known as the bell curve.

Graphing Frequency Distributions

Line Graphs

I hope you're ready to do some graphing, because *I* am. Would you like to try your hand at drawing a line graph, using the data from Table 4.2? If you don't remember exactly how to do this, please reread the section on line graphs near the beginning of Chapter 1. Then take out a piece of graph paper and start drawing. When you've finished, please compare your work with mine in Figure 4.1.

Figure 4.1: Frequency Distribution of Hourly Wages at the Friendly Financial Corporation, December 19, 2038.

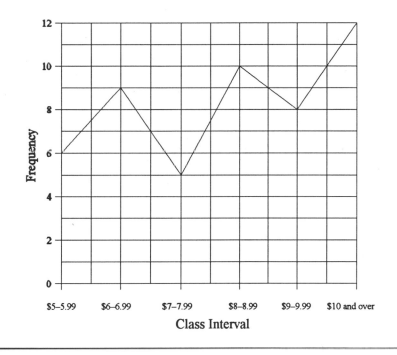

Bar Graphs

Vertical Bar Graphs

You know what? I don't think that that line graph really does that much for our frequency distribution. So I'll tell you what. Let's do a bar graph for the data in Table 4.5. Again, if you don't remember how to do bar graphs, please reread that section in Chapter 1 and then try. After you've drawn your graph, please check your work by looking at Figure 4.2.

Figure 4.2: Frequency Distribution of IQs at the New England School for the Elite, September 18, 2041

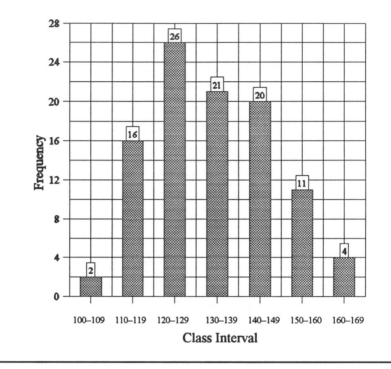

Did your bar graph look anything like mine? OK, what do you think? Do you think bar graphs do a better job than line graphs of depicting frequency distributions?

You'll notice that I wrote the number of each frequency above the corresponding bar in Figure 4.2. Although you certainly don't have to do this, there are two good reasons you should. One, it helps someone looking at your graph to notice immediately exactly how large each frequency is. Two, writing these numbers is a check on your own work. Every so often you'll catch a mistake in your graph this way.

Now I'm going to give you an entire frequency distribution problem to work out. First use the data in Table 4.6 to fill in the tally and frequency columns of Table 4.7. After you've checked your work with mine in Table 4.8, please draw a bar graph. Then compare your results with mine in Figure 4.3.

Table 4.6: Average Hours of Sleep per Night of the Chief Executive Officers of the 200 Top Computer Software Firms, 2011

5	8	6	7	7	8	4	8	6	9
7	8	6	6	8	5	7	6	6	10
5	5	8	4	7	7	8	5	9	6
5	8	5	6	8	7	4	7	9	5
6	5	7	5	8	4	7	6	7	8
4	7	5	6	5	7	9	4	6	7
5	6	4	7	6	8	9	7	8	5
4	6	8	7	8	8	5	7	9	5
6	8	5	6	6	8	5	5	4	6
5	5	8	6	4	8	5	8	8	6
7	7	4	8	3	8	6	7	9	4
9	5	8	8	5	6	6	4	6	9
5	5	8	6	9	5	7	5	8	4
4	6	7	5	7	7	6	8	5	9
4	4	8	6	7	5	5	8	6	5
4	6	3	7	6	6	8	5	7	9
6	6	9	5	6	4	8	6	7	9
8	9	4	6	3	7	5	5	6	4
4	6	4	8	8	6	7	5	8	10
5	5	8	5	7	9	4	8	6	7

Table 4.7: Frequency Distribution of Average Hours of Sleep per Night of the Chief Executive Officers of the 200 Top Computer Software Firms, 2011

Average Hours	Tally	Frequency
3		
4		
5		
6		
7		
8		
9		
10		

Table 4.8: Frequency Distribution of Average Hours of Sleep per Night of the Chief Executive Officers of the 200 Top Computer Software Firms, 2011

Average Hours	Tally	Frequency
3	///	3
4	JHT JHT JHT JHT ////	24
5	JHT JHT JHT JHT JHT JHT JHT JHT /	41
6	JHT JHT JHT JHT JHT JHT JHT JHT //	42
7	JHT JHT JHT JHT JHT JHT ///	33
8	JHT JHT JHT JHT JHT JHT JHT ////	39
9	JHT JHT JHT /	16
10	//	2
		200

Figure 4.3: Frequency Distribution of Average Hours of Sleep per Night of the Chief Executive Officers of the 200 Top Computer Software Firms, 2011

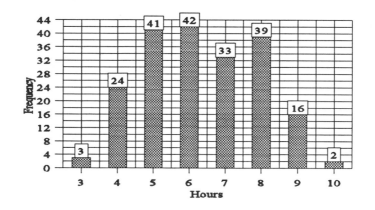

Horizontal Bar Graphs

So far we've done vertical bar graphs. What would our bar graph in Figure 4.3 look like if it were drawn horizontally? It would look like the one I drew in Figure 4.4. Now it's your turn to draw a horizontal bar graph. Use the data in Table 4.9 to fill in the tally and frequency columns of Table 4.10. Compare your work with mine in Table 4.11, and then see if you can draw a horizontal bar graph. Then compare your graph with mine in Figure 4.5.

Figure 4.4: Frequency Distribution of Average Hours of Sleep per Night of the Chief Executive Officers of the 200 Top Computer Software Firms, 2011

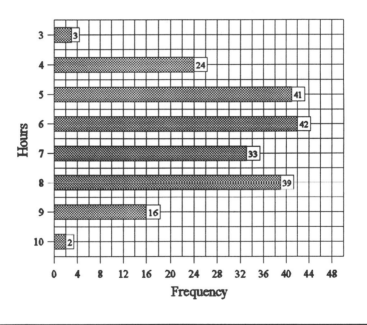

Table 4.9: Course Enrollment at Memphis University, Fall Semester, 2029, 8 A.M. Monday–Wednesday–Friday.

22	31	19	25	29	16	12	36	143	18
72	9	22	35	118	6	15	31	24	27
30	15	153	8	19	30	34	18	7	19
30	26	33	17	75	13	26	10	35	9
16	188	15	32	29	22	31	20	31	14
19	25	90	14	7	78	35	29	23	18
9	66	12	17	33	19	16	23	30	62
28	8	13	17	26	33	45	10	22	17
30	97	11	22	15	27	31	30		

Table 4.10: Frequency Distribution of Course Enrollment at Memphis University, Fall Semester, 2029, 8 A.M. Monday–Wednesday–Friday.

Class Interval	Tally	Frequency
0–9		
10–19		
20–29		
30–39		
40 and over		

Table 4.11: Frequency Distribution of Course Enrollment at Memphis University, Fall Semester, 2029, 8 A.M. Monday–Wednesday–Friday.

Class Interval	Tally	Frequency
0–9	ЖÍ ///	8
10–19	ЖÍ ЖÍ ЖÍ ЖÍ ЖÍ ///	28
20–29	ЖÍ ЖÍ ЖÍ ЖÍ	20
30–39	ЖÍ ЖÍ ЖÍ ЖÍ	20
40 and over	ЖÍ ЖÍ //	12
		88

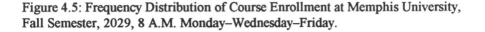

Figure 4.5: Frequency Distribution of Course Enrollment at Memphis University, Fall Semester, 2029, 8 A.M. Monday–Wednesday–Friday.

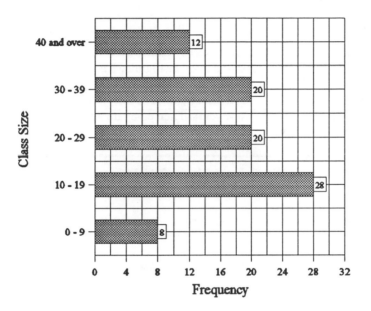

Histograms

Toward the end of Chapter 1 we drew some histograms. Let's do one more. This time we'll work a little less formally.

Jinx and Dana Adams ran a market research firm based in Springfield, Massachusetts. Their firm was hired by a real estate developer who was interested in building an 80–lane bowling alley in downtown Springfield. Using the data below, draw up a frequency distribution, using intervals of 0, 1, 2, 3, 4, and over 4 times per month.

Responses to question: If a large bowling alley were built in downtown Springfield, how many times a month would you bowl?

1, 0, 1, 0, 0, 4, 1, 0, 3, 6, 1, 0, 0, 4, 0, 2, 5, 0, 0, 0, 4, 2, 1, 1, 0, 8, 3, 0, 0,
3, 0, 1, 0, 0, 2, 4, 0, 0, 4, 0, 1, 1, 4, 0, 4, 0, 3, 2, 0, 0, 0, 1, 0, 0, 3, 1, 0, 10,
2, 0, 1, 1, 0, 4, 0, 6, 0, 3, 0, 4, 0, 1, 0, 0, 8, 1, 2, 0, 4, 1, 4, 0, 0, 0, 0, 0, 5, 0,
1, 1, 4, 0, 10, 6, 0, 0, 3, 1, 4, 2, 1, 4, 0, 2, 0, 1, 1, 4, 4, 0, 2, 1, 2

After you've compared your frequency distribution with mine in Table 4.12, go ahead and draw a frequency distribution histogram. Since a histogram looks like a vertical bar graph that's been pushed together, I haven't bothered to draw one. But if you don't remember what a histogram looks like, just go back to Chapter 1, where you'll find a few, and draw yours. Then compare it to mine in Figure 4.6.

Table 4.12: Frequency Distribution of Springfield, Massachusetts, Bowling Survey

Times per Month	Tally	Frequency
0	JHT JHT JHT JHT JHT JHT JHT JHT JHT II	47
1	JHT JHT JHT JHT III	23
2	JHT JHT	10
3	JHT II	7
4	JHT JHT JHT I	16
Over 4	JHT IIII	9

Figure 4.6: Frequency Distribution of Springfield, Massachusetts, Bowling Survey

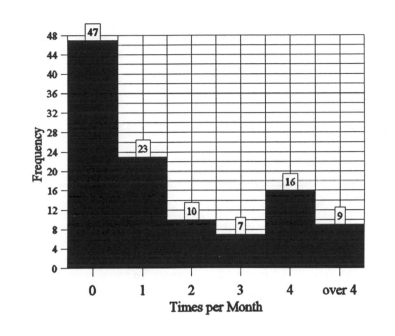

Cumulative Frequency

One more thing to cover and we're outta here. Cumulative frequency is one of the easiest topics in the entire book. We find it by doing a running total of the frequencies.

That's just what I've done in Table 4.13. The cumulative frequency of all the numbers in the top two class intervals is 15 (6 + 9). The cumulative frequency of all the numbers in the top three class intervals is 20 (6 + 9 + 5, or 15 + 5). You get the picture.

Table 4.13: Frequency Distribution of Hourly Wages Paid to Employees at the Friendly Financial Corporation, December 19, 2038

Class Interval	Frequency	Cumulative Frequency
$ 5.00–$5.99	6	6
6.00– 6.99	9	15
7.00–7.99	5	20
8.00–8.99	10	30
9.00–9.99	8	38
10.00 and over	12	50

Cumulative frequencies tell us how many numbers are below a certain value. The cumulative value of 30 tells us that 30 employees earned less than $9. And a cumulative frequency of 15 tells us that 15 employees earned less than $7.

Well, it's time for *you* to find some cumulative frequencies. Please fill in the cumulative frequency column in Table 4.14, and then check your work by glancing at Table 4.15.

Table 4.14: Frequency Distribution of IQs at the New England School for the Elite, September 18, 2041

Class Interval	Frequency	Cumulative Frequency
100–109	2	
110–119	16	
120–129	26	
130–139	21	
140–149	20	
150–159	11	
160–169	4	

Table 4.15: Frequency Distribution of IQs at the New England School for the Elite, September 18, 2041

Class Interval	Frequency	Cumulative Frequency
100–109	2	2
110–119	16	18
120–129	26	44
130–139	21	65
140–149	20	85
150–159	11	96
160–169	4	100

If you got that right then you can skip the next problem.

Problem: Please fill in the cumulative frequencies in Table 4.16 and then compare your work to mine in Table 4.17.

Table 4.16: Frequency Distribution of Average Hours of Sleep per Night of the Chief Executive Officers of the 200 Top Computer Software Firms, 2011

Average Hours	Frequency	Cumulative Frequency
3	3	
4	24	
5	41	
6	42	
7	33	
8	39	
9	16	
10	2	

Table 4.17: Frequency Distribution of Average Hours of Sleep per Night of the Chief Executive Officers of the 200 Top Computer Software Firms, 2011

Average Hours	Frequency	Cumulative Frequency
3	3	3
4	24	27
5	41	68
6	42	110
7	33	143
8	39	182
9	16	198
10	2	200

Conclusion

We won't be drawing any more graphs for a while, but the skills you've developed transposing data from tables to graphs will be useful not just in school or on the job, but in helping you to understand much of the data presentation in newspaper and magazine articles. Frequency distribution itself is central to the study of percentiles, which we'll be taking up in Chapter 6. If you're not entirely comfortable doing frequency distribution problems, I urge you to reread at least parts of this chapter before you go on. Remember that learning statistics is a lot like putting up a building. If the foundation is weak, you know what will happen to that building.

Statistics, then, needs to be learned in a certain sequence, because you'll need to master certain concepts, or tools, before you can understand more advanced concepts and tools. In the next chapter, for example, we'll work with percentages, which we'll need when we begin percentiles in the chapter after that. And when we finally get to the normal distribution and the bell curve in Chapter 8, we'll be drawing upon our knowledge of percentiles.

Learning statistics on your own can be an extremely difficult task. So please don't move on to any new material until you fully understand the preceding material. By that I mean you're getting a high percentage of the problems right. And speaking of percentages, that's exactly what comes next.

Chapter 5

Percents

Back in Chapter 2 we changed fractions into decimals, and now we'll be changing decimals into percents. What *is* a percent? A percent is most often the result of expressing a number as a fraction of 100. Thus, 40% is 40 parts out of 100 parts. And 79% is 79 parts out of 100 parts.

Consider the fraction 1/4, or one-quarter. To change 1/4 into a decimal, we divide 4 into 1 and get .25. A quarter, as we know, is 25 cents. Since a dollar has 100 cents, a quarter is 25/100 of a dollar, which is 25 parts out of 100 parts, or 25%.

In this chapter we'll be changing decimals into percents, changing fractions into percents, finding percentage changes, and, near the end of the chapter, working on percentage distributions. These skills are basic to statistics, and you will find them quite useful in later chapters.

Changing Decimals into Percents

Let's convert .53 into a percent. This number can be thought of as 53 parts out of 100 parts, or 53%. Similarly, .82 can be thought of as 82 parts out of 100 parts, or 82%.

If we step back and look at what we just did, we would see that we simply moved the decimal point two places to the right and tacked on a percentage sign. We started with .53, which we can write as .530 (since adding a zero to a decimal does not change its value), moved its decimal two places to the right,

$$.53.0,$$
$$\overset{\rightarrow}{}$$

and added a percentage sign, 53.0% (any whole number may have a decimal point placed to its right followed by zeros, and its value will remain the same). So to change a decimal into a percent, just move the decimal two places to the right and tack on a percentage sign. And that's *it!*

Here's a problem set to do:

Problems: Convert each of these decimals into percents:

(a) .05   (b) .95   (c) .50   (d) .008   (e) .67

Solutions:

(a) 5%   (b) 95%   (c) 50%   (d) 0.8%   (e) 67%

Now we'll add a new dimension. Convert the number 5.9 into a percent.

What did you get? Did you get 590%? What we do is add a zero to 5.9 to make it 5.90 (you may always add zeros to a decimal without changing its value), move the decimal two places to the right, and add the percent sign. May we add a zero to the number 40 without changing its value? If we add a zero, we get 400. Does 40 = 400? If you think it does, then I'd like to trade my $40 for your $400.

OK, here comes another problem set.

Problems: Convert each of these numbers into percents:

(a) 4.0   (b) 16.1   (c) 40.7   (d) 100.0   (e) 837.9

Solutions:

(a) 400%   (b) 1,600%   (c) 4,070%   (d) 10,000%   (e) 83,790%

If you got those right, then you're ready for another trick. Convert the number 8 into a percent.

Did you get 800%? Here's what we did. We started with 8, added a decimal point and a couple of zeros: 8 = 8.00. Then we converted 8.00 into a percent by moving the decimal two places to the right and adding a percent sign:

$$8.00 = 8.\underrightarrow{00}.\% = 800\%$$

Now we come to another problem set:

Problems: Convert each of these numbers into percents:

(a) 6   (b) 2   (c) 14   (d) 90   (e) 10

Solutions:

(a) 600%   (b) 200%   (c) 1,400%   (d) 9,000%   (e) 1,000%

Changing Fractions into Percents

Remember how we described percents at the beginning of the chapter? A percent is most often the result of expressing a number as a fraction of 100. When we convert fractions into percents, it's like shooting fish in a barrel to convert fractions whose denominators are 100. For instance, convert 87/100 into a percent.

The answer is 87%. What we did was drop the 100 and tack on a percentage sign. Here's a set of problems that are just like that.

Problems: Convert each of these numbers into a percent:

(a) 14/100   (b) 37/100   (c) 9/100   (d) 72/100

Solutions:

(a) 14%   (b) 37%   (c) 9%   (d) 72%

What if our fraction doesn't have a denominator of 100? Very often we can work around that by making the denominator into 100. There's an old law of arithmetic: You may multiply the numerator and the denominator of a fraction by the same number without changing the value of that fraction. See if you can convert 19/50 into a percent.

Solution: $\dfrac{19 \times 2}{50 \times 2} = \dfrac{38}{100} = 38\%$

Now convert 3/4 into a percent:

Solution: $\dfrac{3 \times 25}{4 \times 25} = \dfrac{75}{100} = 75\%$

Can you do this with *every* fraction? No. You can't do it with 17/23, or 8/9, or 14/39. You need to look at the denominator to see if it can be divided evenly into 100. The fractions in the next problem set all have denominators that can be divided evenly into 100.

Problems: Convert each of these fractions into percents.

(a) 1/5   (b) 3/10   (c) 19/20   (d) 1/4

Solutions:

(a) $\dfrac{1 \times 20}{5 \times 20} = \dfrac{20}{100} = 20\%$

(b) $\dfrac{3 \times 10}{10 \times 10} = \dfrac{30}{100} = 30\%$

(c) $\dfrac{19 \times 5}{20 \times 5} = \dfrac{95}{100} = 95\%$

(d) $\dfrac{1 \times 25}{4 \times 25} = \dfrac{25}{100} = 25\%$

Now we're ready to tackle those fractions whose denominators cannot be divided evenly into 100. How do we convert 17/23 into a percent? We follow a three-step process: (1) divide the denominator into the numerator; (2) move the decimal two places to the right; and (3) add a percentage sign. See what you can do with 17/23.

Solution:

$$
\begin{array}{r}
.7391 \ = 73.9\% \\
23\overline{)\ 17.0000} \\
-16\,1\text{xxx} \\
\hline
90 \\
-69 \\
\hline
210 \\
-207 \\
\hline
30 \\
-23 \\
\end{array}
$$

I rounded my answer to one decimal place. For the rest of the problems in this chapter, we'll round to one decimal place, unless otherwise instructed. Now see if you can convert 8/9 into a percent.

Solution:

$$
\begin{array}{r}
.8\ 8\ 8\ 8 \\
9\overline{)\,8.0^80^80^80} = 88.9\%
\end{array}
$$

You probably noticed that I did not use a calculator to do the division. Although you may use a calculator to solve these problems and most of the problems that follow throughout the book, I would suggest that you solve as many as possible without using a calculator. When you're not routinely doing math in your head or with a pencil and paper, your facility with numbers will get quite rusty. Just remember the adage "Use it or lose it."

Problems: Convert each of these problems into percents.

(a) 14/39   (b) 12/23   (c) 7/43   (d) 3/7

Solutions:

(a)

$$
\begin{array}{r}
.3589 \ = 35.9\% \\
39\overline{)\ 14.0000} \\
-11\,7\text{xxx} \\
\hline
2\ 30 \\
-195 \\
\hline
350 \\
-312 \\
\hline
380 \\
-351 \\
\end{array}
$$

(b)

$$\begin{array}{r} .5217 = 52.2\% \\ 23\overline{)12.0000} \\ \underline{-11\ 5xxx} \\ 50 \\ \underline{-46} \\ 40 \\ \underline{-23} \\ 170 \\ \underline{-161} \end{array}$$

(c)

$$\begin{array}{r} .1627 = 16.3\% \\ 43\overline{)7.0000} \\ \underline{-4\ 3xxx} \\ 270 \\ \underline{-2\ 58} \\ 120 \\ \underline{-86} \\ 340 \\ \underline{-301} \end{array}$$

(d)

$$\begin{array}{r} .4\ 2\ 8\ 5 = 42.9\% \\ 7\overline{)3.0^{2}0^{6}0^{4}0} \end{array}$$

Changing Percents into Decimals

These problems should be quite easy for you. Just reverse the process you carried out when you changed decimals into percent. Convert 61.9% into a decimal.

Solution: 61.9% = .619

When we converted decimals into percents we moved the decimal point two places to the right and added a percentage sign. Now we do the exact opposite: We move the decimal point two places to the left and drop the percentage sign.

Problems: Change each of these percents into decimals:

(a) 56.2%   (b) 34.9%   (c) 90.0%   (d) 31%   (e) 8%

Solutions:

(a) .562   (b) .349   (c) .9 (or .90, or .900)   (d)   .31   (e) .08

This last one was a little tricky. If you got it right, then move on to the next section. If not, then please do this next problem set.

Problems: Change each of these percents into decimals.

(a) 6%   (b) 60%   (c) 1%   (d) 4%

   Solutions:

(a) .06   (b) .6 (or .60)   (c) .01   (d) .04

Finding Percentage Changes

Most college graduates cannot figure out percentage changes. You can prove this statement by attending any college graduation in the country and then asking ten graduates selected at random to do a few of the problems from this section. Most of these freshly minted graduates will get most or all of these problems wrong. But when you have completed this section, you will be able to do percentage changes in your sleep.

We're going to start you out with a very easy one. You were staying in a hotel that charged $100 a night and the price was raised to $115. By what percentage did the price go up?

Solution: $\dfrac{\$15}{\$100} = 15\%$

Here's another problem: You were earning $200 a week and you received a $40 pay increase. By what percentage did your salary go up?

Solution: $\dfrac{\$40}{\$200} = \dfrac{\$20}{\$100} = 20\%$

Actually, we applied another old law of arithmetic: You may divide the numerator and the denominator of a fraction by the same number without changing the value of that fraction. Suppose Wei Wong charged $20 an hour and raised his hourly rate to $25 an hour. By what percentage did Mr. Wong's hourly rates go up?

Solution: $\dfrac{\$5 \times 5}{\$20 \times 5} = \dfrac{\$25}{\$100} = 25\%$

Problems: For each of these problems, find the percentage change.

(a)   A price is raised from $100 to $138.

(b)   A price is raised from $1,000 to $1,100.

(c)   Attendance rose from $200 to $275.

(d)   Rent rose from $500 to $825.

Solutions:

(a) $\dfrac{\$38}{100} = 38\%$

(b) $\dfrac{\$100}{\$1000} = \dfrac{\$10}{\$100} = 10\%$

(c) $\dfrac{\$75}{\$200} = \dfrac{\$37.50}{\$100} = 37.5\%$

(d) $\dfrac{\$325}{\$500} = \dfrac{\$65}{\$100} = 65\%$

Can we also do percentage decreases this way? Definitely! Find the percentage decrease if we go from 50 to 35.

Solution: $\dfrac{-15}{50} = \dfrac{-30}{100} = -30\%$

Problems: For each of these problems find the percentage change.

(a)   Sally Tierney's auto insurance premium was lowered from $500 to $440.

(b)   Elizabeth Storey's daily highway mileage fell from 100 to 73.

(c)   The strikeouts of the Chicago Cubs pitching staff fell from 1,000 to 932.

(d)   The number of accidents on I-25 during the New Year's weekend fell from 20 to 9.

Solutions:

(a) $\dfrac{-\$60}{\$500} = \dfrac{-\$12}{\$100} = -12\%$

(b) $\dfrac{-27}{100} = -27\%$

(c) $\dfrac{-68}{1000} = \dfrac{-6.8}{100} = -6.8\%$

(d) $\dfrac{-11}{20} = \dfrac{-55}{100} = -55\%$

So far all our problems had denominators that were 100 or could easily be changed to 100. But how do we find percentage changes when we don't have that denominator to work with? For example, how do we find the percentage change in *this* problem: When you stepped on the scale one morning and saw that you weighed 220 pounds, you resolved then and there to lose 40 pounds. If you did, then what percentage of your weight would you have lost?

Solution:

$$\text{Percentage change} = \frac{\text{change}}{\text{original number}} = \frac{40}{220} = \frac{4}{22} = \frac{2}{11}$$

$$
\begin{array}{r}
.1818 = 18.2\% \\
11\overline{)\,2.0000} \\
-11\text{xxx} \\
\hline
90 \\
-88 \\
\hline
20 \\
-11 \\
\hline
90 \\
-88 \\
\hline
\end{array}
$$

Here's another one. Find the percentage change if your earnings rose from $19,500 to $21,000.

$$\text{Percentage change} = \frac{\text{change}}{\text{original number}} = \frac{\$1,500}{\$19,500} = \frac{15}{195} = \frac{3}{39} = \frac{1}{13}$$

$$
\begin{array}{r}
.0769 = 7.7\% \\
13\overline{)\,1.0000} \\
-91\text{xx} \\
\hline
90 \\
-78 \\
\hline
120 \\
-117 \\
\hline
\end{array}
$$

Problems:

(a)   Luigi Antonelli raised his SAT score from 950 to 1,080 by taking an SAT preparation course. By what percentage did his score rise?

(b)   Lars Andersen lowered his cholesterol level from 209 to 173. By what percentage did his cholesterol level decline?

(c)   Kyra Markova's real estate taxes rose from $6,000 to $8,500. By what percentage did they rise?

(d)   Harriet Gold's time for running a mile fell from 11 minutes to 8 minutes. By what percentage did her time fall?

Solutions:

(a)   Percentage change $= \dfrac{\text{change}}{\text{original number}} = \dfrac{130}{950} = \dfrac{13}{95}$

$$
\begin{array}{r}
.1368 = 13.7\% \\
95\overline{)13.0000} \\
-9\,5\text{xxx} \\
\hline
3\ 50 \\
-2\ 85 \\
\hline
650 \\
-570 \\
\hline
800 \\
-760 \\
\end{array}
$$

(b)   Percentage change $= \dfrac{\text{change}}{\text{original number}} = \dfrac{36}{209}$

$$
\begin{array}{r}
.1722 = 17.2\% \\
209\overline{)36.0000} \\
-20\ 9\text{xxx} \\
\hline
15\ 10 \\
-14\ 63 \\
\hline
470 \\
-418 \\
\hline
520 \\
-418 \\
\end{array}
$$

(c)   Percentage change $= \dfrac{\text{change}}{\text{original number}} = \dfrac{\$2,500}{\$6,000} = \dfrac{25}{60} = \dfrac{5}{12}$

$$
\begin{array}{r}
.4166 = 41.7\% \\
12\overline{)5.0000} \\
-4\ 8\text{xxx} \\
\hline
20 \\
-12 \\
\hline
80 \\
-72 \\
\hline
80 \\
-72 \\
\end{array}
$$

(d)   Percentage change = $\dfrac{\text{change}}{\text{original number}} = \dfrac{3}{11}$

$$
\begin{array}{r}
.2727 = 27.3\% \\
11\overline{)\,3.0000} \\
\underline{-2\,2\text{xxx}} \\
80 \\
\underline{-77} \\
30 \\
\underline{-22} \\
80 \\
\underline{-77}
\end{array}
$$

Percentage Distribution

Finding Percentage Distributions

We'll start with a ridiculously easy question: If a room contained half men and half women, what percentage of the people in the room were women and what percent were men? Obviously, the room had 50% women and 50% men. That's all there is to percentage distribution. Of course, the problems do get a bit more complicated, but all percentage distribution problems start out with one simple fact: The distribution will always add up to 100%.

Here's another one. One-fifth of the adults in Cairo, Illinois, are unemployed, two-fifths are working part-time, and the rest are working full-time. What is the percentage distribution of Cairo's unemployed, part-time employed, and full-time employed?

Solution:

$$\text{Unemployed} = \frac{1}{5} = \frac{1 \times 20}{5 \times 20} = \frac{20}{100} = 20\%$$

$$\text{Part-time employed} = \frac{2}{5} = 40\%$$

$$\text{Full-time employed} = 1 - \frac{3}{5} = \frac{2}{5} = 40\%$$

Here's another question. If, over the course of a week, you obtained 250 grams of protein from red meat, 150 from fish, 100 from poultry, and 50 from other sources, what percentage of your protein intake came from red meat and what percentage came from each of the other sources?

red meat	250 grams
fish	150 grams
poultry	100 grams
other	50 grams
	550 grams

Try to work this out for yourself. Hint: 550 grams = 100%.

Solution:

$$\text{red meat} = \frac{250}{550} = \frac{25}{55} = \frac{5}{11} = 45.5\% \qquad 11\overline{)5.0^60^50^60}\ ^{.4\ 5\ 4\ 5}$$

$$\text{fish} = \frac{150}{550} = \frac{15}{55} = \frac{3}{11} = 27.3\% \qquad 11\overline{)3.0^80^30^80}\ ^{.2\ 7\ 2\ 7}$$

$$\text{poultry} = \frac{100}{550} = \frac{10}{55} = \frac{2}{11} = 18.2\% \qquad 11\overline{)2.0^90^20^90}\ ^{.1\ 8\ 1\ 8}$$

$$\text{other} = \frac{50}{550} = \frac{5}{55} = \frac{1}{11} = 9.1\% \qquad 11\overline{)1.00^100}\ ^{.09\ 09}$$

Check:

$$\begin{array}{r}
^3 4^1 5.5 \\
2\ 7.3 \\
1\ 8.2 \\
9.1 \\
\hline
1\ 0\ 0.1
\end{array}$$

Sometimes you can save a few steps by taking a shortcut. When you do a lot of work with numbers, you get to recognize shortcuts such as the one shown in the box below.

Numerical Shortcut

When you glance back at the problem we just completed, you'll notice that 5/11, 3/11, 2/11, and 1/11 of your diet came from those four different sources.

To find the percentage distribution we did four separate division operations. But we can actually get away with doing just one.

By dividing 1 by 11, we find that the fraction 1/11 comes to 9.09% (before rounding). We can find the percentages of the other sources by simple multiplication:

$$5/11 = .0909 \times 5 = .4545 = 45.5\%$$

$$3/11 = .0909 \times 3 = .2727 = 27.3\%$$

$$2/11 = .0909 \times 2 = .1818 = 18.2\%$$

When you are using a calculator, this shortcut may not save you much time, but you may not always have one available. In general, as you work your way through the book, see if you can find some shortcuts that even I may have missed.

When doing percentage distribution problems, it's always a good idea to check your work. If your percentages don't add up to 100, then you've definitely made a mistake, so you'll need to go back over all your calculations. Because of rounding, my percentages added up to 100.1. Occasionally, you'll end up with 100.1 or 99.9 when you check, which is fine.

Problems:

(a)   In a psychology class, 4 students are psychology majors, 5 are sociology majors, 6 are business majors, and 10 others have no declared major. What is the percentage share of each major?

(b)   George Stevens consumes 2,200 calories a day. He consumes 400 calories of fat, 500 calories of protein, and the rest in carbohydrates. What is the percentage share of calories of each of the three food groups?

(c)   Eleni Zimiles has 8 red beads, 4 blue beads, 3 white beads, 2 yellow beads, and 1 green bead. What is the percentage distribution of Eleni's beads?

(d)   Georgia Pacific ships 5,000 freight containers a week. Fifteen hundred are sent by air, 2,300 by rail, and the rest by truck. What percentage are sent by air, rail, and truck, respectively?

Solutions:

(a)   Total number of students $= 4 + 5 + 6 + 10 = 25$

psychology majors $= \dfrac{4 \times 4}{25 \times 4} = \dfrac{16}{100} = 16\%$

sociology majors $= \dfrac{5 \times 4}{25 \times 4} = \dfrac{20}{100} = 20\%$

business majors = $\dfrac{6 \times 4}{25 \times 4} = \dfrac{24}{100} = 24\%$

undeclared majors = $\dfrac{10 \times 4}{25 \times 4} = \dfrac{40}{100} = 40\%$

Check:
$$16$$
$$20$$
$$24$$
$$\underline{40}$$
$$100$$

(b) fat = $\dfrac{400}{2200} = \dfrac{4}{22} = \dfrac{2}{11} = \quad$ $\overset{.1\ 8\ 1\ 8}{11\overline{\smash{)}2.0^90^20^80}} = 18.2\%$

protein =

$$\dfrac{500}{2200} = \dfrac{5}{22} = \quad \overset{.2272}{22\overline{\smash{)}5.0000}} = 22.7\%$$
$$\underline{-44\text{xxx}}$$
$$60$$
$$\underline{-44}$$
$$160$$
$$\underline{-154}$$
$$60$$
$$\underline{-44}$$

carbohydrates = $\dfrac{1300}{2200} = \dfrac{13}{22}$ $\quad \overset{.5909}{22\overline{\smash{)}13.0000}} = 59.1\%$
$$\underline{-110\text{xxx}}$$
$$2\,00$$
$$\underline{-1\,98}$$
$$200$$
$$\underline{-198}$$

Check:
$$18.2$$
$$22.7$$
$$\underline{59.1}$$
$$100.0$$

(c) 8
     4
     3
     2
     1
    ––
    18

red $=$ $\dfrac{8}{18} = \dfrac{4}{9}$ $9\overline{)4.0^40^40}$ $\dfrac{.4\ 4\ 4}{} = 44.4\%$

blue $=$ $\dfrac{4}{18} = \dfrac{2}{9}$ $9\overline{)2.0^20^20}$ $\dfrac{.2\ 2\ 2}{} = 22.2\%$

white $=$ $\dfrac{3}{18} = \dfrac{1}{6}$ $6\overline{)1.0^40^40^40}$ $\dfrac{.1\ 6\ 6\ 6}{} = 16.7\%$

yellow $=$ $\dfrac{2}{18} = \dfrac{1}{9}$ $9\overline{)1.0^10^10}$ $\dfrac{.1\ 1\ 1}{} = 11.1\%$

green $=$ $\dfrac{1}{18}$

$$18\overline{)1.0000} \quad .0555 = 5.6\%$$
$$\underline{-90}\text{xx}$$
$$100$$
$$\underline{-90}$$
$$100$$
$$\underline{-90}$$
$$10$$

Check:
$^24^24.4$
$2\ 2.2$
$1\ 6.7$
$1\ 1.1$
$\underline{5.6}$
$1\ 0\ 0.0$

(d) air $=$ $\dfrac{1500}{5000} = \dfrac{15}{50} = \dfrac{30}{100} = 30\%$

     rail $=$ $\dfrac{2300}{5000} = \dfrac{23}{50} = \dfrac{46}{100} = 46\%$

     truck $=$ $\dfrac{1200}{5000} = \dfrac{12}{50} = \dfrac{24}{100} = 24\%$

Check: 30
                 46
                _24
                100

Pie Charts

Percentage distributions can be illustrated by means of bar graphs and histograms, but usually we use pie charts. And the sum of the slices always adds up to the whole pie, or 100%.

We'll start out with a freshly baked, unsliced pie (see Figure 5.1). Suppose we divided it into four equal slices. Each slice would be 25% of the pie. See if you can divide the pie into those four 25% slices, one of which will go to Stu, one to Kyra, one to Jessica, and one to Jane. Go ahead and slice up the pie, labeling each person's slice and percentage share.

Figure 5.1: Pie Chart

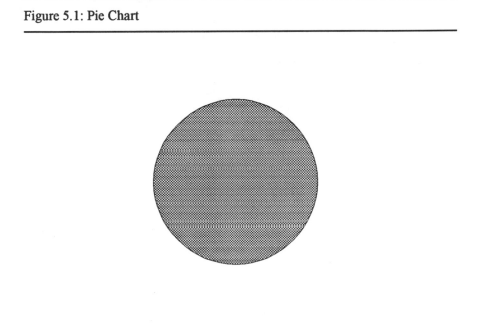

Your pie should look like the one I've sliced up in Figure 5.2. In general, you need to draw your slices in proportion to the percentage shares. Now see what you can do with the data from problem (a) of the last problem set to fill in the empty pie shown in Figure 5.3. Then compare your pie with mine shown in Figure 5.4.

Figure 5.2: Pie Divided in Four Equal Slices

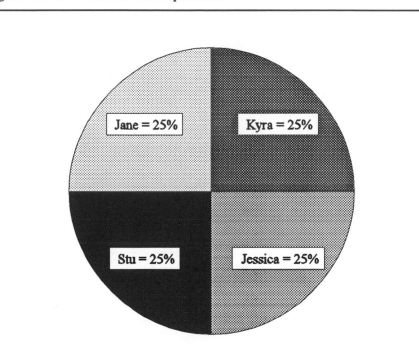

Figure 5.3: Pie Chart to Be Filled In with Data from Problem (a)

Figure 5.4: Pie Chart Representing Student Majors

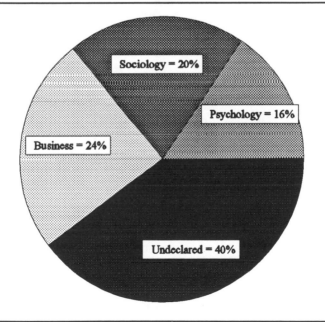

For our next pie-slicing exercise, use the data from problem (b) to fill in Figure 5.5. Then compare your pie chart with mine in Figure 5.6.

Figure 5.5: Pie Chart to Be Filled In with Data from Problem (b)

Figure 5.6: Pie Chart of Calories from Food Groups

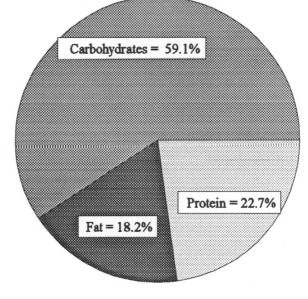

I'll tell you what. Let's do just one more and then we can call it a day. Use the data from problem (c) to fill in Figure 5.7 and then compare your work with mine in Figure 5.8.

Figure 5.7: Pie Chart to Be Filled In with Data from Problem (c)

Figure 5.8: Pie Chart of Beads

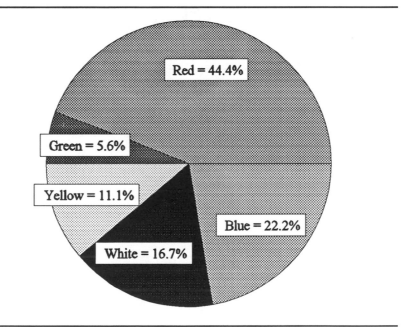

Chapter 6

Percentiles

Now that you've learned everything you'll ever need to know about percents, you may be asking yourself: What's the difference between a percent and a percentile? A percent is most often the result of expressing a number as a fraction of 100. Thus, 60% is 60 parts out of 100 parts.

So what's a percentile? It's the ranking of a score in comparison to other scores. The percentile rank of a score, then, represents the percent of cases in a group that achieved scores *lower* than the one cited. For example, if you got a score of 572 on the verbal part of the SATs, and if that score has a percentile rank of 60, it means that you got a higher score than 60% of the people who took that exam.

Incidentally, each score is considered to be a hypothetical point without dimension, so it would be equally correct to say that 40% of the test takers scored *higher* than 572. Percentile ranks are often expressed to the nearest hundredth. So if your percentile rank was 43.82, that would mean that your score was *higher* than 43.82% of the test takers but *lower* than 56.18% of those who took the test.

The main thing we'll be doing in this chapter is converting scores into percentiles and percentiles into scores. We'll be answering two basic questions: (1) if you know your score, what was your percentile rank and (2) if you know your percentile rank, what was your score? If we happen to have a cumulative percentage distribution, then we have enough information to answer these questions.

Using a Cumulative Percentage Graph

The easiest way of converting scores into percentile ranks and percentile ranks into scores is by using a cumulative percentage graph, if you just happen to have one. Let's see how you do. Using Figure 6.1, if you earned $60,000, what would your percentile rank be?

Figure 6.1: Cumulative Percentage Graph of Earnings of Employees at Hypothetical ABC Corporation

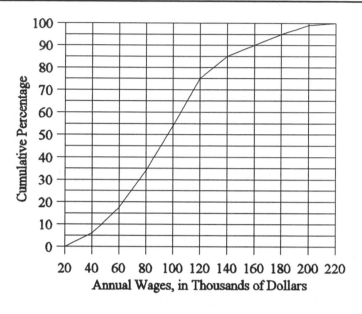

It would be about 17.5. Your earnings would be higher than those of 17.5% of your fellow employees—and lower than 82.5% of them. All you need to do is find annual wages of $60,000 on the horizontal axis, run your finger up to the cumulative distribution curve, and then left across the graph to the vertical axis listing the cumulative percentage.

If you earned $130,000, what would your percentile rank be? Write it down. And also write down the percentage of your fellow employees who would have lower earnings and the percentage that would have higher earnings.

Your earnings would be in the 80th percentile. This means that you would be earning more than 80% of your coworkers, but that 20% of them would be earning more than you.

Now we'll reverse field and see if you can find your earnings if you know your percentile rank. If you were in the 30th percentile, about how much would you be earning?

You would be earning about $74,000. Just move your finger from the 30 on the vertical axis directly to the right until it hits the cumulative percentage distribution curve, and then straight down to the horizontal axis. I see your earnings as about $74,000, give or take maybe $1,000. Here's one more: If your earnings were in the 99th percentile, approximately how much would they be?

They would be approximately $210,000. Because the cumulative percentage distribution curve has flattened out, it's hard to answer with great precision. Any answer within the range of, say, $203,000 to $212,000 would be acceptable.

Deciles and Quartiles

If your score is in the first decile, what does that mean? It means that your score is in the 10th percentile. Remember that the prefix "dec" means "ten." What would your percentile rank be if you were in the 8th decile?

You would be in the 80th percentile. OK, if your score is in the first quartile, what does *that* mean?

It means that your score is in the 25th percentile. What would your percentile rank be if you were in the third quartile?

It would be the 75th percentile. Now try *this* one on for size: If your score were the median score, in what decile and in what quartile would you be?

You would be in the fifth decile and the second quartile. Quartiles and deciles are terms used occasionally in statistics, almost always in association with percentiles, but I wouldn't lose any sleep over them.

Finding Scores, Percentiles, Deciles, and Quartiles on the Cumulative Percentage Graph

We'll start off with an easy one. If you were in the second decile in Figure 6.2, how much would your batting average be?

Figure 6.2: Major League Batting Averages for 2026 Season

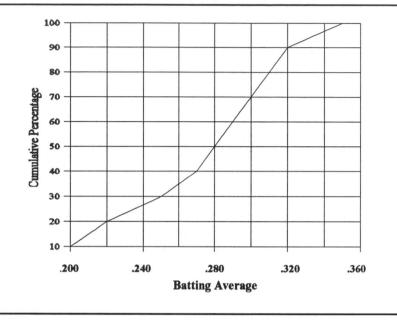

Your batting average would be .220. Incidentally, if you're not up on batting averages, a batting average of .220 means that the hitter gets hits 22% of the time she or he comes to bat. A hitter batting .340 gets 34 hits every 100 times at bat. What would your batting average be if you were in the third quartile?

Your batting average would be about .306. Now we'll start with the batting average. If you were batting .280, what decile would you be in?

You'd be in the fifth decile. Also, incidentally, you would have the mean batting average. And if you were batting .234, what quartile would you be in?

You'd be in the first quartile. Now let's try something a little more complex. If you were batting .289, what decile would you be in and what would your percentile rank be?

You would be in the sixth decile and the 60th percentile rank. Last question: If you were batting .348, what would be your decile, quartile, and percentile rank?

You would be in the tenth decile, the fourth quartile, and the hundredth percentile. We're assuming that you have the highest batting average in the major leagues. To be in the hundredth percentile, you would need to have a higher score than anyone else.

You may have noticed that Figure 6.2 is not a complete cumulative percentage graph since it begins at a cumulative percentage of 10. Had we gone down to zero to show the cumulative percentages of batting averages from .000 to .200, Figure

6.2 would have to have been drawn more than twice as wide. As we noted back in Chapter 1, we sometimes need to draw truncated graphs.

Obtaining Percentile Ranks from Frequency Distribution Tables

So far we've been finding percentile ranks from frequency distribution graphs. But these graphs are not always available. Furthermore, we may need greater precision than is possible with graphical representation. So what do we do? We go back to using the same type of frequency distribution tables we compiled near the end of Chapter 4.

We are going to use Table 6.1 to find scores corresponding to given percentile ranks. We are also going to find the percentile ranks of various scores. Back in Chapter 3 we went over the process of interpolation. Now we shall have a chance to apply that process.

Table 6.1: Frequency Distribution of Results of Hypothetical IQ Test Given to Random Sample of Ivy League Freshmen, 2022

Class Interval	f	Cumulative f
170–179	2	1,434
160–169	10	1,432
150–159	86	1,422
140–149	231	1,336
130–139	493	1,105
120–129	475	612
110–119	94	137
100–109	37	43
90–99	6	6

Let's find the percentile rank of someone with an IQ of 122. Looking at the cumulative frequency column of Table 6.1, we can see that 137 people had IQs of less than 120. Since 1,434 people were tested, an IQ of 120 would be in the 9.55th percentile (137/1434 = .0955). And someone with an IQ of 129 would be in the 42.68th percentile (612/1434 = .4268).

Remember that a score's percentile rank tells us that score is higher than a certain percentage of people who took the same test. Since a total of 1,434 people were tested and we see that 137 got scores lower than 120, the percentile rank of a score of 120 would be 9.55, or 9.55%. Similarly, since 612 people got scores below 130, then the percentile rank of a score of 130 would be 42.68, or 42.68%.

But what's the percentile rank of someone with an IQ of 122? To find that percentile rank we will need to do some interpolation. We know it would be between 9.55 and 42.68, but it would be much closer to 9.55.

I know you are very eager to do that interpolation, but first we need to adjust our class intervals. You see, the true limits of the class interval 120 to 129 is really 119.5 to 129.5. Suppose that someone's computed IQ came to 119.7. It would be rounded up to 120. And if it were 129.3, it would be rounded down to 129. So when you think about it, IQs rounded to 120 include those between 119.5 and 120.5. And IQs rounded to 129 include those between 128.5 and 129.5. So let's redo Table 6.1 and 6.2, adding a column of true limits for class intervals.

Table 6.2: Frequency Distribution of Results of Hypothetical IQ Test Given to Random Sample of Ivy League Freshmen, 2022

Class Interval	True Limits	f	Cumulative f
170–179	169.5–179.5	2	1434
160–169	159.5–169.5	10	1432
150–159	149.5–159.5	86	1422
140–149	139.5–149.5	231	1336
130–139	129.5–139.5	493	1105
120–129	119.5–129.5	475	612
110–119	109.5–119.5	94	137
100–109	99.5–109.5	37	43
90–99	89.5–99.5	6	6

Now we can interpolate. Let's restate the two percentile rankings we did earlier. The percentile rank of an IQ of 119.5 is 9.55, and the percentile rank of an IQ of 129.5 is 42.68. We need to find the percentile rank of an IQ of 122.

Using the true limits, each class interval has a range of 10. The one we shall use, of course, is from 119.5 to 129.5. Where is 122 located within this range? It is a distance of 2.5 from 119.5, or 2.5/10 the distance through the entire range. In other words, an IQ of 122 is .25, or 25% through the range of 119.5–129.5.

We already found in Table 6.1 that 137 people have IQs below 119.5. And we also know that there are 475 people who have IQs between 119.5 and 129.5. Assuming they are evenly distributed throughout the range, we can further assume that 25% of them lie below 122. So to find the percentile rank of an IQ of 122, all we need to do is add 25% of 475 to 137 and then convert that number into a percentile rank.

475 × .25 = 118.75. Adding 118.75 to 137, we find a total of 255.75 IQs below 122. We can convert this score into a percentile by dividing it by the total number of people tested, 1434. 255.75/1434 = 17.83. So an IQ of 122 would be in the 17.83rd percentile.

What would be the percentile rank for a person with an IQ of 155? Use Table 6.2.

An IQ of 155 lies within the class interval with true limits of 149.5–159.5. We see that 1,336 people had IQs of less than 149.5, and there are 86 people in the 149.5 – 159.5 range. An IQ of 155 is 5.5/10 of the way through that range. 86 × .55 = 47.4. We add these 47.4 scores to the 1,336 scores below 149.5 to get 1383.4. 1383.4/1,434 = 96.47th percentile.

Using Table 6.2 we can also obtain scores at different percentiles. First we'll find the score at the 70th percentile (or seventh decile).

Before we can find that IQ, we'll need to find the cumulative frequency corresponding to that score. We can do that by applying this formula:

$$\text{cumulative } f = \frac{\text{percentile rank} \times n}{100}$$

$$\text{cumulative } f = \frac{70 \times 1434}{100}$$

$$\text{cumulative } f = \frac{10,038}{100}$$

$$\text{cumulative } f = 1,003.8$$

A cumulative frequency of 1,003.8 is just under 1,105, which is the total of IQs below 139.5. So the IQ we're looking for is also just below 139.5.

There are 493 scores within the range of 129.5 – 139.5. Our score is only 1.2 from the top of the range, or 491.8 from the bottom. In other words, it is 491.8/493 of the way through the range. This translates to 99.76% of the way through the range. Since the range is 10, 10 × .9976 = 9.976. We add that to the IQ at the bottom of the range of 129.5 to get an IQ of 139.476. This, incidentally, rounds to 139.5.

Next problem: Find the median IQ (which is also the IQ at the fifth decile, second quartile, or 50th percentile).

Solution:

$$\text{cumulative } f = \frac{\text{percentile rank} \times n}{100}$$

$$\text{cumulative } f = \frac{50 \times 1434}{100}$$

$$\text{cumulative f} = \frac{71,700}{100}$$

$$\text{cumulative f} = \quad 717$$

A cumulative frequency of 707 places us fairly close to the low end of the interval with the true limits of 129.5 – 139.5. There are 493 IQs in the range of this interval. If there are 612 IQs below 129.5, then a cumulative frequency of 707 is 95 scores into the interval, or 95/493 of the way through the interval. This comes to 19.27% of the way through the interval. .1927 × 10 = 1.927. We add 1.927 to 129.5 to get an IQ of 131.427.

Here's a two-part problem based on the data in Table 6.3. What are the percentile ranks of weights of (a) 59.5 and (b) 119.5? You'll find that this type of problem is easier to solve than the ones we've been doing.

Table 6.3: Frequency Distribution of Results of Hypothetical Sampling of the Weights of Sixth Graders

Class Interval	True Limits	f	Cumulative f
140–159	139.5–159.5	4	293
120–139	119.5–139.5	33	289
100–119	99.5–119.5	84	256
80–99	79.5–99.5	109	172
60–79	59.5–79.5	56	63
40–59	39.5–59.5	7	7

Solutions:

(a) $\dfrac{7}{293}$ = .02389 = 2.39th percentile

(b) $\dfrac{256}{293}$ = .8737 = 87.37th percentile

What made this two-part problem so easy, of course, was that there was no interpolation. That situation *is* unusual, but it does come up from time to time. Now I'd like you to solve this two-part problem. What are the percentile ranks of weights of (a) 130 pounds and (b) 85 pounds?

Solutions:

(a) $\dfrac{10.5}{20} = .525$ $.525 \times 33 = 17.325$

$17.325 + 256 = 273.325$ $\dfrac{273.325}{293} = .9328$

93.28th percentile

(b) $\dfrac{5.5}{20} = .275$ $.275 \times 109 = 29.975$

$29.975 + 63 = 92.975$ $\dfrac{92.975}{293} = .3173$

31.73rd percentile

Now we'll do a set of problems where we start with the percentile and find the score. Find the weight of a person in the (a) second quartile; (b) sixth decile; (c) 15th percentile; (d) 58th percentile.

Solutions:

(a)   cumulative f $= \dfrac{\text{percentile rank} \times \text{n}}{100}$

$= \dfrac{50 \times 293}{100}$

$= \dfrac{14{,}650}{100}$

$= 146.5$

146.5 is in 79.5–99.5 interval.

weight $= 79.5 + 67/109 \times 20$

weight $= 79.5 + .6147 \times 20$

weight $= 79.5 + 12.294$

weight $= 91.794$ pounds

(b)   cumulative f $= \dfrac{\text{percentile rank} \times \text{n}}{100}$

$= \dfrac{60 \times 293}{100}$

$= \dfrac{17{,}580}{100}$

$= \quad 175.8$

175.8 is in 99.5–119.5 interval.

weight $\quad = \quad 99.5 + 3.8/84 \times 20$

weight $\quad = \quad 99.5 + .0452 \times 20$

weight $\quad = \quad 99.5 + .904$

weight $\quad = \quad 100.404$ pounds

(c)   cumulative f $\quad = \quad \dfrac{\text{percentile rank} \times n}{100}$

$= \quad \dfrac{15 \times 293}{100}$

$= \quad \dfrac{4395}{100}$

$= \quad 43.95$

43.95 is in the interval 59.5–79.5.

weight $\quad = \quad 59.5 + 36.95/56 \times 20$

weight $\quad = \quad 59.5 + .6598 \times 20$

weight $\quad = \quad 59.5 + 13.196$

weight $\quad = \quad 72.696$ pounds

(d)   cumulative f $\quad = \quad \dfrac{\text{percentile rank} \times n}{100}$

$= \quad \dfrac{58 \times 293}{100}$

$= \quad \dfrac{16,994}{100}$

$= \quad 169.94$

169.94 is in the interval 79.5–99.5.

weight $\quad = \quad 79.5 + 106.94/109 \times 20$

weight $\quad = \quad 79.5 + .9811 \times 20$

weight $\quad = \quad 79.5 + 19.622$

weight $\quad = \quad 99.122$ pounds

Finding the Cumulative Frequency and Setting True Limits

All of the problems you've worked out so far have been set up for you with cumulative frequencies and true limits. Now it's time for you to set them up yourself. Use the data in Table 6.4 to set up the cumulative frequency and true limits. Then check your work with mine in Table 6.5.

Table 6.4: Frequency Distribution of Items Sold at Happy Harry's Discount Store, January 8, 2033

Class Interval	f
$.90–.99	3
.80–.89	8
.70–.79	20
.60–.69	16
.50–.59	7
.40–.49	2

Table 6.5: Frequency Distribution of Items Sold at Happy Harry's Discount Store, January 8, 2033

Class Interval	True Limits	f	Cumulative f
.90–.99	$.895–.995	3	56
.80–.89	.795–.895	8	53
.70–.79	.695–.795	20	45
.60–.69	.595–.695	16	25
.50–.59	.495–.595	7	9
.40–.49	.395–.495	2	2

We don't want to let a perfectly good table go unused. (1) Please find the percentile ranks of items sold for (a) 73 cents; (b) 56 cents. (2) Please find how much items are sold for with percentile ranks of (a) 92; (b) 20.

Solutions:

(1)   (a) $\dfrac{35}{10} = .35$ $35 \times 20 = 7$

                 $7 + 25 = 32$ $\dfrac{32}{56} = \dfrac{4}{7} = .5714$

                          57.14th percentile

       (b) $\dfrac{6.5}{10} = .65$ $.65 \times 7 = 4.55$

                 $4.55 + 2 = 6.55$ $\dfrac{6.55}{56} = .1167$

                          11.67th percentile

(2)   (a)   cumulative $f = \dfrac{\text{percentile rank} \times n}{100}$

                          $= \dfrac{92 \times 56}{100}$

                          $= \dfrac{5152}{100}$

                          $= 51.52$

       51.52 is in the 79.5 – 89.5 interval.

                 price $= 79.5 + \dfrac{6.52}{8} \times 10$

                 price $= 79.5 + .815 \times 10$

                 price $= 79.5 + 8.15$

                 price $= 87.65$ cents

       (b)   cumulative $f = \dfrac{\text{percentile rank} \times n}{100}$

                          $= \dfrac{20 \times 56}{100}$

                          $= \dfrac{1120}{100}$

                          $= 11.2$

11.2 is in the 59.5 – 69.5 interval.

$$\text{price} = 59.5 + \frac{2.2}{16} \times 10$$

$$\text{price} = 59.5 + .1375 \times 10$$

$$\text{price} = 59.5 + 1.375$$

$$\text{price} = 60.875 \text{ cents}$$

Now that I've gotten you to find the true limits of a frequency distribution, I'm going to ask you a strange question: Do we always need to use true limits? Until now you may have gotten the impression that the answer is yes. But when class limits are relatively large, the distinction between these intervals and true limits becomes trivial. There's a really absurd example illustrated by Table 6.6, which will show you just how trivial the distinction between class intervals and true limits can be.

Table 6.6: Class Intervals of Earnings of the Top Executives of the Super Software Corporation

Class Interval
0–$999,999
$1,000,000–$1,999,999
2,000,000–2,999,999
3,000,000–3,999,999
4,000,000–4,999,999
5,000,000–5,999,999

What are the true limits for these intervals? Work them out right on Table 6.6, and then check your work with mine in Table 6.7.

Table 6.7: Class Intervals of Earnings of the Top Executives of the Super Software Corporation

Class Interval	True Limits
0–$999,999	0–$999,999.49
$1,000,000–$1,999,999	$ 999,999.50–$1,999,999.49
2,000,000–2,999,999	2,999,999.50–2,999,999.49
3,000,000–3,999,999	3,999,999.50–3,999,999.49
4,000,000–4,999,999	4,999,999.50–4,999,999.49
5,000,000–5,999,999	5,999,999.50–5,999,999.49

Would it make any sense to do true limits for the class intervals shown in Tables 6.6 and 6.7? Obviously not. As I've indicated, the difference between class intervals and corresponding true limits is trivial. How trivial? Let's compare the ranges of the first class interval and its true limits.

The range of the class interval is $999,999 and the range of its true limits is $999,999.49. And the range of each of the remaining class intervals is $999,999, while the ranges of the remaining true limits are $1,000,000.

OK, so we don't use true limits. But when we do our computations converting scores into percentiles and percentiles into scores, do we use class intervals of $999,999? That would make for some pretty unwieldy computations. So just round off the range of the class interval to $1,000,000.

Is there a clear dividing line between when we need to use true limits and when we can just use our class intervals? Unfortunately, there isn't any clear dividing line. We saw that it is absurd to use true limits when the class intervals are relatively large. We certainly would not want to bother using true limits when we don't need a great deal of precision. Suppose, for example, we needed to find just an approximate percentile or an approximate score. Clearly, the class interval would do the job. Whether we use true limits is a matter of judgment, which comes with additional experience.

Let's do one more set of problems, this time *not* using true limits. Use the data in Table 6.8 to solve these problems. Be sure, however, to add a column for cumulative frequency.

Table 6.8: Frequency Distribution of Salaries at the Jessica Storey Corporation, 2033

Class Interval	f
$120,000–$139,999	2
100,000–119,999	51
80,000–99,999	126
60,000–79,999	290
40,000–59,999	411
20,000–39,999	273
0–19,999	95

Here are a couple of problem sets based on the data in Table 6.8. (1) What are the percentile ranks of salaries of (a) $115,000; (b) $80,000; and (c) $22,765? (2) Find the salaries of people in (a) the third quartile; (b) the fourth decile; (c) the 91st percentile; and (d) the 14th percentile.

Solutions:

Table 6.9: Cumulative Frequency Distribution of Salaries at the Jessica Storey Corporation, 2033

Class Interval	f	cumulative f
$120,000–$139,999	2	1,248
100,000–119,999	51	1,246
80,000–99,999	126	1,195
60,000–79,999	290	1,069
40,000–59,999	411	779
20,000–39,999	273	368
0–19,999	95	95

(1)   (a) $\dfrac{\$15,000}{\$20,000} = \dfrac{3}{4} = .75$ $.75 \times 51 = 38.25$

$$38.25 + 1195 = 1233.25$$

$$\frac{1233.25}{1248} = .9882 \quad 98.82\text{nd percentile}$$

(b) $\quad \dfrac{1069}{1248} = .8566 = 85.66\text{th percentile}$

(c) $\quad \dfrac{\$2765}{\$20,000} = .13825 \quad .13825 \times 273 = 37.743$

$$37.7423 + 95 = 132.7423$$

$$\frac{132.7423}{1248} = .1064 \quad 10.64\text{th percentile}$$

(2)   (a)   cumulative f $= \dfrac{\text{percentile rank} \times \text{n}}{100}$

$$= \frac{75 \times 1248}{100}$$

$$= \frac{93,600}{100}$$

$$= 936$$

936 is in the \$60,000 – \$79,999 interval.

salary $= \$60,000 + \dfrac{157}{290} \times \$20,000$

salary $= \$60,000 + .54138 \times \$20,000$

salary $= \$60,000 + \$10,827.60$

salary $= \$70,827.60$

(b)   cumulative f $= \dfrac{\text{percentile rank} \times n}{100}$

$= \dfrac{40 \times 1248}{100}$

$= \dfrac{49,920}{100}$

$= 499.20$

499.2 is in the \$40,000 – \$59,999 interval.

salary $=$ \$40,000 $+ \dfrac{131.2}{411} \times$ \$20,000

salary $=$ \$40,000 $+ .3192 \times$ \$20,000

salary $=$ \$40,000 $+$ \$6,384

salary $=$ \$46,384

(c)   cumulative f $= \dfrac{\text{percentile rank} \times n}{100}$

$= \dfrac{91 \times 1248}{100}$

$= \dfrac{113,568}{100}$

$= 1,135.68$

1,135.68 is in the \$80,000 – \$99,999 interval.

salary $=$ \$80,000 $+ \dfrac{66.68}{126} \times$ \$20,000

salary $=$ \$80,000 $+ .5292 \times$ \$20,000

salary $=$ \$80,000 $+$ \$10,584

salary $=$ \$90,584

(d)   cumulative f $= \dfrac{\text{percentile rank} \times n}{100}$

$= \dfrac{14 \times 1248}{100}$

$= \dfrac{17{,}472}{100}$

$= 174.72$

174.72 is in the \$20,000 – \$39,999 interval.

salary $= \$20{,}000 + \dfrac{79.72}{273} \times \$20{,}000$

salary $= \$20{,}000 + .292 \times \$20{,}000$

salary $= \$20{,}000 + \$5{,}840$

salary $= \$25{,}840$

Percentiles are a very useful statistical tool that you'll probably employ from time to time. After we cover standard deviation in the next chapter, you'll get a chance to employ percentiles quite extensively in the chapter on the normal curve.

Chapter 7

Standard Deviation

Have you ever attended any meetings of the National Association for the Advancement of Fat Americans (NAAFA)? The mean, or average, weight of the attendees at the last meeting was, say, 315 pounds. Suppose a man weighing 350 arrived a few minutes late. What would he do to the average weight in the room?

Clearly, he would raise it. Now, do you think everyone would be saying to themselves, "Wow! Is that guy *fat*"? Not likely.

OK, how fat *is* this guy? Or, more to the point we'll be making, would we consider a weight of 350 somewhat distant from the mean of 315?

What if all the people in the room had weights clustered between 300 and 330? Then this late arrival would be the heaviest person there. On the other hand, if there were people who weighed 200, 300, 400, and 500 pounds, then his weight would be pretty close to the average.

The big question, then, is how widely dispersed are the scores (or weights) about the mean? The dispersion measure statisticians use the most is the standard deviation.

Finding the standard deviation is as bad as it gets—at least in *this* book. You'll need to do a lot of calculations. But there's one bit of good news: You can do all those calculations on a pocket calculator.

We'll be using two basic formulas—one for ungrouped data and the other for grouped data (from a frequency distribution). Each of these formulas may at first seem intimidating, but you'll get used to working with them and grow to love them. Would you believe, maybe, *like* them?

We're still not ready to define just what the standard deviation is, but you may be able to make a guess about its magnitude. Remember that it measures dispersion of scores about the mean. So, if we have a large standard deviation, say, 40 pounds, what does that mean?

It means that the weights are fairly widely dispersed. And if we have a relatively small standard deviation, say, 15 pounds?

Then the weights are very closely clustered about the mean.

Here is an array of weights at the NAAFA meeting: 297, 301, 306, 312, 314, 317, 325, 329, 334, and 350.

Before we find the standard deviation, I'd like you to find the new mean for the numbers in this array.

You should have gotten 318.5. So this latecomer raised the mean weight of the NAAFA meeting's attendees from 315 to 318.5.

Now let's get down to cases. We'll start by finding the standard deviation of ungrouped data, like the array we've been discussing. Then, in the second part of the chapter we'll move on to finding the standard deviation of grouped data.

Calculating Standard Deviation with Ungrouped Scores

How much is the standard deviation of weights at the NAAFA meeting? To find out we follow our familiar three-step process: (1) Write down a formula; (2) substitute numbers into the formula; and (3) solve for the unknown. Here's our formula.

$$\text{standard deviation} = \sqrt{\frac{N \sum X^2 - (\sum X)^2}{N(N-1)}}$$

We'll go over each of the terms here. The symbol that encloses everything else, is a square root sign. It means you need to find the square root of everything within the sign after you've done all the other stuff you're supposed to do. But don't worry; you'll be able to find the square root in about three seconds with your calculator.

In case you don't remember the other terms, N is the number of scores in the array. The Greek letter sigma, or $\sum$, means "the sum of." Here we'll be summing the squares of each of the scores in the array. The next term, $(\sum X)^2$, is often misused. We sum up all the scores, and then we square that sum. Got all that? Well, don't worry, because I'm going to work out this problem step by step.

First do a vertical array of X's like those in Table 7.1a.

Table 7.1a

X
297
301
306
312
314
317
325
329
334
350

Next, you'll need to add up the X's, which gives you $\sum X$. Then do another column a bit to the right of the X column for X^2s in Table 7.1b.

Table 7.1b

X	X^2
297	88,209
301	90,601
306	93,636
312	97,344
314	98,596
317	100,489
325	105,625
329	108,241
334	111,556
350	122,500
$\sum X = 3,185$	$\sum X^2 = 1,016,797$

We've gone ahead and added the X^2s. Now we're ready to substitute numbers into the standard deviation formula. First, how much is N?

N is 10 since there are 10 scores in the array. The sum of the X^2s, or $\sum X^2$ is 1,016,797.

What about $(\Sigma X)^2$? How do we find that number? What do you think?

We just square the sum of the X's, or multiply $3,185 \times 3,185$. That gives us 10,144,225.

Finally, for the denominator, $N(N - 1)$, we substitute $10(10 - 1)$.

$$\sqrt{\frac{N\Sigma X^2 - (\Sigma X)^2}{N(N-1)}} = \sqrt{\frac{10(1,106,797) - 10,144,225}{10(10-1)}}$$

$$\sqrt{\frac{10,167,970 - 10,144,225}{10(9)}} = \sqrt{\frac{23,745}{90}}$$

$$\sqrt{263.8333} = 16.24$$

Do you need to memorize this formula? Not unless you're a masochist.

Someday you may actually need to find a standard deviation. If you've still got this book—or any other statistics book—you'll be able to look up this formula. Remember, the trick is not to be able to memorize formulas, but to be able to use them.

You're going to be getting a lot of practice. Let's go back to our NAAFA meeting and alter the weight distribution.

If their weights are 180, 213, 264, 291, 307, 316, 339, 350, 430, and 495, let's find the standard deviation. Do you think it will be larger or smaller than our previous standard deviation?

Because the weights are much more dispersed about the mean, which, incidentally, is still 318.5, the standard deviation will be much larger. We find the ΣX and the ΣX^2 first (see Table 7.2).

Table 7.2

X	X^2
180	32,400
213	45,369
264	69,696
291	84,681
307	94,249
316	99,856
339	114,921
350	122,500
430	184,900
495	245,025
$\Sigma X = 3,185$	$\Sigma X^2 = 1,093,597$

$$\text{standard deviation} = \sqrt{\frac{N\sum X^2 - (\sum X)^2}{N(N-1)}} = \sqrt{\frac{10(1,093,597) - (3185)^2}{10(10-1)}}$$

$$= \sqrt{\frac{10,935,970 - 10,144,225}{10(9)}} = \sqrt{\frac{781,645}{90}}$$

$$= \sqrt{8684.944} = 93.19$$

Clearly this distribution is much more dispersed about the mean than the previous one. After all, we've got a standard deviation of 93.19 pounds here compared to just 16.24 pounds in the earlier distribution.

Are you beginning to get the impression that just maybe something is missing from this picture? So far we've said that the standard deviation measures the degree of dispersion about the mean. And that the larger the standard deviation, the larger the degree of dispersion. So far, so good. But what exactly *is* the standard deviation?

I *could* give you a definition that you'll find less than satisfying: The standard deviation is the square root of the sum of squared deviations from the mean divided by N – 1. But if you're willing to wait until we get to the next chapter, when we discuss the normal curve, I promise you a much more practical definition. But if you just can't wait, then see the box below.

What Is the Standard Deviation?

This concept can be explained by example. Suppose there were four people, aged 10, 15, 18, and 25. Their average age is 17.

How much do each of their ages deviate from the mean? The deviations of their ages from the mean would be –7, –2, +1, and +8. Now square these deviations and add the squares: 49 + 4 + 1 + 64 = 118. Divide 118 by the number of terms, 4, and get 29.5. The square root of 29.5, or 5.43 is the standard deviation.

Now we're going to work out some standard deviation problems.

Problem 1: Calculate the standard deviation for this set of measurements: 2, 4, 7, 7, 10, 13. Then find the mean.

Remember to set up a table with a column of Xs and a column of X^2s, and then to plug your $\sum$Xs and your $\sum X^2$s into the standard deviation formula and solve for the standard deviation. After you've done that, check your work with my solution below.

Solution to problem 1:

Table 7.3

X	X²
2	4
4	16
7	49
7	49
10	100
13	169
$\sum X = 43$	$\sum X^2 = 387$

standard deviation $= \sqrt{\dfrac{N\sum X^2 - (\sum X)^2}{N(N-1)}} = \sqrt{\dfrac{6(387) - (43)^2}{6(6-1)}}$

$= \sqrt{\dfrac{2322 - 1849}{6(5)}} = \sqrt{\dfrac{473}{30}}$

$= \sqrt{15.7666} = 3.97$

$\overline{X} = \dfrac{\sum X}{N} = \dfrac{43}{6} = 7.17$

Problem 2: Calculate the standard deviation for this set of measurements: 1, 6, 8, 12, 14, 15, 20. Then find the mean.

Solution to problem 2:

Table 7.4

X	X²
1	1
6	36
8	64
12	144
14	196
15	225
20	400
$\sum X = 76$	$\sum X^2 = 1{,}066$

standard deviation $= \sqrt{\dfrac{N\sum X^2 - (\sum X)^2}{N(N-1)}} = \sqrt{\dfrac{7(1066) - (76)^2}{7(7-1)}}$

$$= \sqrt{\dfrac{7462 - 5776}{7(6)}} = \sqrt{\dfrac{1686}{42}}$$

$$= \sqrt{40.14286} = 6.34$$

$$\overline{X} = \dfrac{\sum X}{N} = \dfrac{76}{7} = 10.86$$

Problem 3: Calculate the standard deviation for this set of measurements: 25, 28, 31, 36, 43, 50, 58, 68. Then find the mean.

Solution to problem 3:

Table 7.5

X	X²
25	625
28	784
31	961
36	1,296
43	1,849
50	2,500
58	3,364
68	4,624
$\sum X = 339$	$\sum X^2 = 16,003$

$$\text{standard deviation} = \sqrt{\frac{N \sum X^2 - (\sum X)^2}{N(N-1)}} = \sqrt{\frac{8(16,003) - (339)^2}{8(8-1)}}$$

$$= \sqrt{\frac{128,024 - 114,921}{8(7)}} = \sqrt{\frac{13,103}{56}}$$

$$= \sqrt{233.98214} = 15.30$$

$$\overline{X} = \frac{\sum X}{N} = \frac{339}{8} = 42.38$$

Problem 4: Calculate the standard deviation for this set of measurements: 4, 4, 4, 4, 4, 4, 4, 4, 4, 4. Is this a trick question? Yes and no. Work it out and then check your work against my solution.

Solution to problem 4:

Table 7.6

X	X²
4	16
4	16
4	16
4	16
4	16
4	16
4	16
4	16
4	16
4	16
$\sum X = 40$	$\sum X^2 = 160$

$$\text{standard deviation} = \sqrt{\frac{N \sum X^2 - (\sum X)^2}{N(N-1)}} = \sqrt{\frac{10(160) - (40)^2}{10(10-1)}}$$

$$= \sqrt{\frac{1600 - 1600}{10(9)}} = \sqrt{\frac{0}{90}} = 0$$

Why was the standard deviation zero? Because there was no dispersion about the mean. The mean was 4 and each number in the distribution was 4.

For the rest of the chapter we'll be finding the standard deviations of grouped data. We'll get to use a new formula with even more complex calculations. As I've said, it doesn't get any harder than this. Once we've worked our way through this second part of the chapter, the rest of the book will be all downhill.

Calculating Standard Deviation with Grouped Scores

Data is not always available in individual measurements, and even if it were, it would sometimes be unwieldy to use for standard deviation calculations. For example, suppose we had 100 people whose ages varied from a few months old to 93. Imagine adding up their ages and then adding the squares of their ages.

Let's get right into it. We'll be finding the standard deviation of the frequency distribution shown in Table 7.7.

Table 7.7

Class Intervals	f	Midpoint of Interval
18–20	1	
15–17	3	
12–14	4	
9–11	5	
6–8	2	
3–5	1	

Now let's get to work. We'll need to find several numbers that will go into the formula we'll be using. First find the total number of measurements, N. And then find the midpoints of each interval.

Have you done that? Then check your answers with mine in Table 7.8.

Table 7.8

Class Intervals	f	Midpoint of Interval	X^2
18–20	1	19	
15–17	3	16	
12–14	4	13	
9–11	5	10	
6–8	2	7	
3–5	1	4	
	N = 16		

And now, in Table 7.8, please fill in the X^2 column. Just square each Midpoint of Interval X.

Check your answers with mine in Table 7.9. In Table 7.9 fill in the column labeled fX. After you've done that, add up these numbers to get $\sum fX$.

Table 7.9

Class Interval	f	Midpoint of Interval	X^2	fX	fX^2
18–20	1	19	361		
15–17	3	16	256		
12–14	4	13	169		
9–11	5	10	100		
6–8	2	7	49		
3–5	1	4	16		
	N = 16				

Check your answers with mine in Table 7.10. Now fill in the column fX^2 (by multiplying the numbers in the f column and those in the X^2 column). When you've done that, add up the numbers in the fX^2 column to get $\sum fX^2$.

Table 7.10

Class Interval	f	Midpoint of Interval	X^2	fX	fX^2
18–20	1	19	361	19	
15–17	3	16	256	48	
12–14	4	13	169	52	
9–11	5	10	100	50	
6–8	2	7	49	14	
3–5	1	4	16	4	
	N = 16			$\sum fX = 187$	

If everything came out right, then you should have the same numbers that I have in Table 7.11.

Table 7.11

Class Interval	f	Midpoint of Interval	X^2	fX	fX^2
18–20	1	19	361	19	361
15–17	3	16	256	48	968
12–14	4	13	169	52	676
9–11	5	10	100	50	500
6–8	2	7	49	14	98
3–5	1	4	16	4	16
	N = 16			$\Sigma fX = 187$	$\Sigma fX^2 = 2,619$

We'll be plugging these numbers into our standard deviation formula for grouped data:

$$\text{standard deviation} = \sqrt{\frac{N \Sigma fX^2 - (\Sigma fX)^2}{N(N-1)}}$$

See if you can do that yourself. Then check my work below:

$$= \sqrt{\frac{16(2619) - (187)^2}{16(16-1)}}$$

And finally, we calculate the standard deviation:

$$= \sqrt{\frac{41,904 - 34,969}{16(15)}} = \sqrt{\frac{6,935}{240}}$$

$$= \sqrt{28.8958} = 5.38$$

To find the standard deviation we need to find the fX and the fX^2 from the table, plug these numbers into the standard deviation formula, and solve for the standard deviation.

Incidentally, how much is the mean? See if you can work it out.

Solution: $$\overline{X} = \frac{\Sigma fX}{N} = \frac{187}{6} = 11.69$$

This frequency distribution has a mean of 11.69 and a standard deviation of 5.38. You'll notice that the scores are somewhat clustered about the mean.

Let's find the mean and standard deviation of another frequency distribution that has more widely dispersed scores (see Table 7.12). Before we begin our calculations, do you expect a larger or smaller standard deviation than in our last problem?

Table 7.12

Class Interval	f	Midpoint of Interval	X^2	fX	fX^2
18–20	4				
15–17	3				
12–14	1				
9–11	2				
6–8	3				
3–5	3				

I know I said it would be a larger standard deviation. Let's find out how much larger. First fill in Table 7.12. Then check your work with mine in Table 7.13. After that we'll take the data we've worked up and plug it into the standard deviation formula.

Table 7.13

Class Interval	f	Midpoint of Interval	X^2	fX	fX^2
18–20	4	19	361	76	1,444
15–17	3	16	256	48	768
12–14	1	13	169	13	169
9–11	2	10	100	20	200
6–8	2	7	49	21	147
3–5	3	4	16	12	48
	N = 16			$\sum fX = 190$	$\sum fX^2 = 2{,}776$

How did you do? If your data checks out with mine, then see if you can calculate the standard deviation. If it didn't, then go back over your work to find out what you did wrong. Once you have, then go on to finding the standard deviation.

standard deviation $= \sqrt{\dfrac{N\sum fX^2 - (\sum fX)^2}{N(N-1)}} = \sqrt{\dfrac{16(2776) - (190)^2}{16(16-1)}}$

$$= \sqrt{\dfrac{44,416 - 36,100}{16(15)}} = \sqrt{\dfrac{8,316}{240}}$$

$$= \sqrt{34.65} = 5.89$$

Now I'd like you to calculate the mean for this frequency distribution.

Solution: $\overline{X} = \dfrac{\sum fX}{N} = \dfrac{190}{16} = 11.88$

Guess what we're going to do next. You guessed it! We're going to do a set of problems very much like the ones we've just done.

Problem 1. Find the mean and standard deviation for the frequency distribution shown in Table 7.14.

Table 7.14

Class Interval	f	Midpoint of Interval	X^2	fX	fX^2
31–35	2				
26–30	4				
21–25	7				
16–20	9				
11–15	8				
6–10	5				
1–5	3				

Table 7.15

Class Interval	f	Midpoint of Interval	X^2	fX	fX^2
31–35	2	33	1089	66	2178
26–30	4	28	784	112	3136
21–25	7	23	529	161	3703
16–20	9	18	324	162	2916
11–15	8	13	169	104	1352
6–10	5	8	64	40	320
1–5	3	3	9	9	27
	N = 38			$\Sigma fX = 654$	$\Sigma fX^2 = 13,632$

standard deviation $= \sqrt{\dfrac{N \Sigma fX^2 - (\Sigma fX)^2}{N(N-1)}} = \sqrt{\dfrac{38(13,632) - (654)^2}{38(38-1)}}$

$\qquad\qquad = \sqrt{\dfrac{518,016 - 427,716}{38(37)}} = \sqrt{\dfrac{90,300}{1406}}$

$\qquad\qquad = \sqrt{64.22475} = 8.01$

$$\overline{X} = \frac{\Sigma fX}{N} = \frac{654}{38} = 17.21$$

Problem 2. Find the mean and standard deviation for the frequency distribution shown in Table 7.16.

Table 7.16

Class Interval	f	Midpoint of Interval	X^2	fX	fX^2
56–60	6				
51–55	17				
46–50	29				
41–45	22				
36–40	14				
31–35	8				

Table 7.17

Class Interval	f	Midpoint of Interval	X^2	fX	fX^2
56–60	6	58	3364	348	20,184
51–55	17	53	2809	901	47,753
46–50	29	48	2304	1392	66,816
41–45	22	43	1849	946	40,678
36–40	14	38	1444	532	20,216
31–35	8	33	1089	264	8,712
	N = 96			$\Sigma fX = 4,383$	$\Sigma fX^2 = 204,359$

$$\text{standard deviation} = \sqrt{\frac{N \Sigma fX^2 - (\Sigma fX)^2}{N(N-1)}} = \sqrt{\frac{96(204,359) - (4,383)^2}{96(96-1)}}$$

$$= \sqrt{\frac{19,618,464 - 19,210,689}{96(95)}} = \sqrt{\frac{407,775}{9120}}$$

$$= \sqrt{44.71217} = 6.69$$

$$\overline{X} = \frac{\Sigma fX}{N} = \frac{4,383}{96} = 45.66$$

Problem 3. We saved the best, or at least the hardest, for last. Be very careful figuring out the midpoints for each class interval. They do not come out to whole numbers. In fact you'll find that each ends in a .5.

Table 7.18

Class Interval	f	Midpoint of Interval	X^2	fX	fX^2
70–79	4				
60–69	9				
50–59	16				
40–49	20				
30–39	17				
20–29	12				
10–19	7				
0–9	1				

Solution:

Table 7.19

Class Interval	f	Midpoint of Interval	X^2	fX	fX^2
70–79	4	74.5	550.25	298	22,201.00
60–69	9	64.5	4160.25	580.5	37,442.25
50–59	16	54.5	2970.25	872	47,524.00
40–49	20	44.5	1980.25	890	39,605.00
30–39	17	34.5	1190.25	586.5	20,234.25
20–29	12	24.5	600.25	294	7,203.00
10–19	7	14.5	210.25	101.5	1,471.75
0–9	1	4.5	20.25	4.5	20.25
	N = 86			$\sum fX$=3,627.0	$\sum fX^2$=175,701.50

$$\text{standard deviation} = \sqrt{\frac{N \sum fX^2 - (\sum fX)^2}{N(N-1)}} = \sqrt{\frac{86(175,701.5) - (3,627)^2}{86(86-1)}}$$

$$= \sqrt{\frac{15,110,329 - 13,155,129}{86(85)}} = \sqrt{\frac{1,955,200}{7310}}$$

$$= \sqrt{267.46922} = 16.35$$

$$\overline{X} = \frac{\sum fX}{N} = \frac{3,627}{86} = 42.17$$

Chapter 8

The Normal Curve

On the first day of the semester, someone in the class can always be counted upon to ask, "Do you grade on a curve?" If you do, then probably a grade of 56 would be passing.

Now we're going to be a lot more specific. When we talk about grading on a curve, we are talking about the normal curve, often referred to as the bell curve. All of the graphs in this chapter, starting with Figure 8.1, are graphs of normal curves. Why are they called bell curves? Take a look at Figure 8.1 and then you tell me.

Figure 8.1

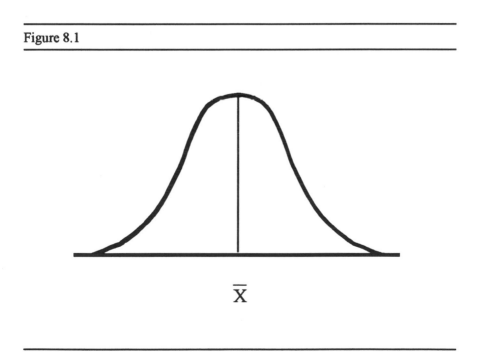

$$\overline{X}$$

I hope you said that the curve in Figure 8.1 is shaped like a bell. To be fair, only the middle of the curve is shaped like a bell. The curve also has tails, extending to the left and the right.

Back in 1995 Richard Herrnstein and Charles Murray immortalized this curve with their best-seller, *The Bell Curve.** The main thesis of the book is that IQ (intelligence quotient) is the most important factor in determining our earnings. So the smarter you are, the more you are likely to earn. Although this book was highly controversial, few people actually read it, perhaps because it is 872 pages long and somewhat technical. But after you've worked your way through this chapter, *The Bell Curve* will be a piece of cake.

I've gotten myself so worked up about *The Bell Curve* that I've completely forgotten about whether a grade of 56 would be passing. You'd probably call it a cop-out if I said that it all depended on the curve used in making up grades. Actually we need to see how the normal curve works before we can use it for grading or for any other purpose. So let's get that out of the way and then we'll see just how good that 56 really is.

The Standard Normal Distribution

The mean is literally the central score in a normal distribution. In Figure 8.1 the mean, $\bar{X}$, is smack in the middle of the curve. Since half the scores are above the mean and half are below it, the mean and the median are identical.

The normal curve is marked off by standard deviations. Since the mean is exactly at the center of the curve, it is 0 standard deviations away from the center. If we move one standard deviation to the left of the mean, we are −1 standard deviation from the mean. This is shown in Figure 8.2. Similarly if we move one standard deviation to the right of the mean, we are +1 standard deviation from the mean. If we move farther to the right, we'll reach 2 standard deviations and 3 standard deviations. And if we move farther to the left from the mean, we would reach −1, −2, and −3 standard deviations.

---

*Published by Simon and Schuster, New York.

Figure 8.2

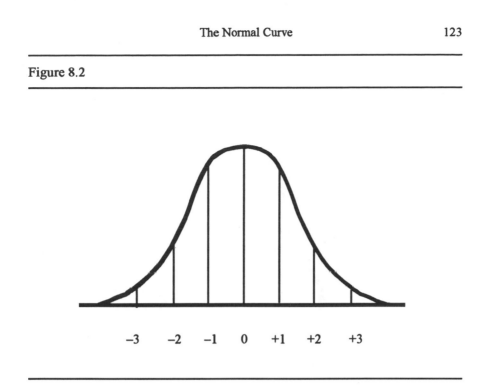

$$-3 \quad -2 \quad -1 \quad 0 \quad +1 \quad +2 \quad +3$$

What does all of this *mean*? I hope Figure 8.3 will help answer that question. You'll notice that the normal curve is divided into sections bounded by standard deviations. Each section is assigned a proportion, or percentage. For example, the section between –1 standard deviations and 0 standard deviations has a percentage of 34.13. This means that in a normal distribution, 34.13 percent of the measurements, or observations, would occur between –1 standard deviations and 0 standard deviations. And 13.59 percent of all observations would be between –1 standard deviations and –2 standard deviations. Since 2.15 percent of the observations would lie between –2 standard deviations and –3 standard deviations, and 0.13 percent of all observations would lie to the left of –3 standard deviations, then what is the total percentage of observations that lie to the left of the mean?

Figure 8.3

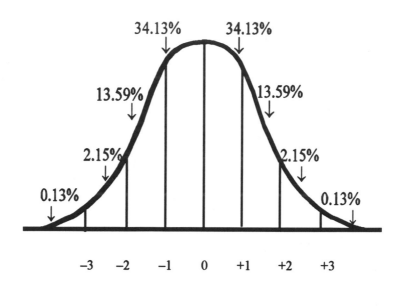

Fifty percent of all observations lie to the left of the mean. And since the standard normal curve is symmetrical, the sections bordered by the positive standard deviations are identical to those on the left half of the curve. So 50 percent of the area under the curve is to the left of the mean, and 50 percent of the area under the curve is to the right of the mean.

Suppose we had a score, or observation, that was equal to the mean. What would be the percentile rank of that score? It would be in the 50th percentile. Ready for another one? Suppose someone got a really low score, say at exactly –2 standard deviations. See if you can figure out its percentile rank.

It would be in the 2.28th percentile (0.13% + 2.15%). In other words, it would be higher than just 2.28 percent of the scores. And what percent of scores would it be lower than?

It would be lower than 97.72 percent of all scores. Let's try one more. Suppose you got a score at the +1 standard deviation? What would your percentile rank be?

You would be in the 84.13rd percentile (50% + 34.13%). All of this is summarized in Figure 8.4.

Figure 8.4

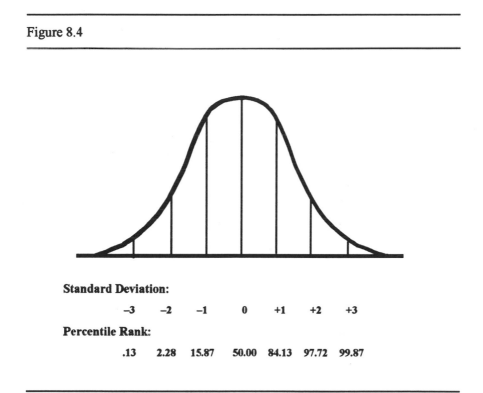

Standard Deviation:

| | −3 | −2 | −1 | 0 | +1 | +2 | +3 |

Percentile Rank:

| | .13 | 2.28 | 15.87 | 50.00 | 84.13 | 97.72 | 99.87 |

You may recall that we never really defined the standard deviation in the last chapter. Now we're equipped to do a better job because of how it's depicted in the standard normal curve. We've already said that the standard deviation is a measure of dispersion of scores about the mean. Let's be more specific. In a normal distribution, 68.26 percent of all scores will lie within one standard deviation of the mean; 95.34 percent of all scores will lie within two standard deviations of the mean; and 99.74 percent of all scores will lie within three standard deviations of the mean.

Not all data falls into perfectly symmetrical normal distributions. Very often curves depicting this data tail off to the left or to the right. We term such curves skewed, and talk about them in the box on the following pages.

Skewed Curves

If we had a standard normal distribution, the mean, median, and mode would be the same number. It would be at the exact center of this symmetrical curve. But some distributions are skewed, such as those shown in Figures 8.5 and 8.6.

Figure 8.5

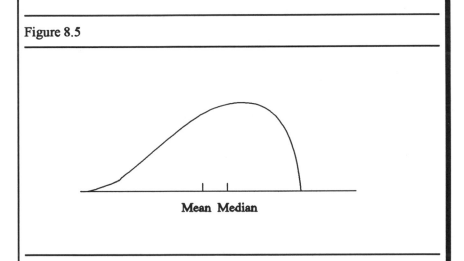

Mean Median

The curve in Figure 8.5 is negatively skewed because it tails off to the left. You'll notice that the mean is lower than the median because it is pulled down by a few extreme low scores. Similarly, you'll notice that the curve in Figure 8.6 is positively skewed because it tails off to the right. The mean is greater than the median because it is pulled up by a few extreme high scores.

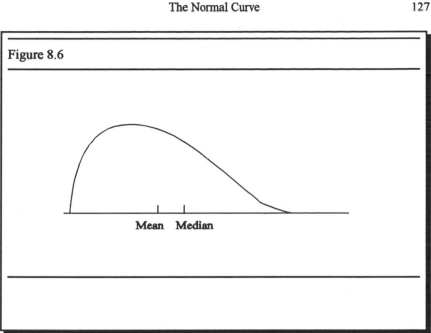

Figure 8.6

Mean   Median

Finding the Percentile Rank

Now we're going to put that standard deviation to work to find out just how well your niece did on that science exam when she received a score of 56. Let's assume the mean grade was 50 and the standard deviation was 6. Can you find her percentile rank?

Solution: Did you get a percentile rank of 84.13? It's helpful to draw a sketch representing the relationships in question, as I've done in Figure 8.7.

Figure 8.7

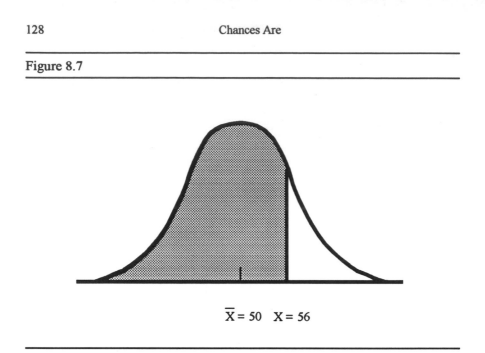

$$\overline{X} = 50 \quad X = 56$$

A grade of 56 would place your niece fairly high in her class. If her instructor marked on a curve, she might give A's to everyone above the 90th percentile, B's to those above the 75th percentile, and lower grades to those who attained lower scores.

Here's another problem. Find the percentile rank of someone who weighs 70 pounds when the standard deviation is 10 and the mean is 90 pounds.

Solution: Again, I've drawn a sketch (see Figure 8.8) to help me see the relationships involved. This person would be in the 2.28th percentile.

Figure 8.8

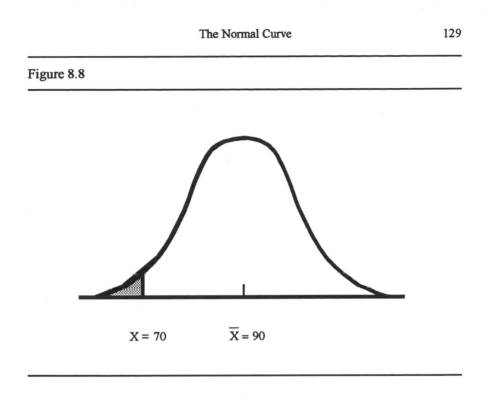

X = 70 $\overline{X}$ = 90

Z-Scores

So far we've used standard deviations of exactly one or two. But most scores fall between the mean and plus or minus one standard deviation, and nearly all the rest are between plus or minus one or two standard deviations from the mean. How do we find the percentile ranks of *these* scores?

What we do is find an intermediate measure called a z-score, and then find our percentile rank, by using Table A in the Appendix at the back of the book.

The z-score is found with this simple formula:

$$z = \frac{X - \overline{X}}{\text{standard deviation}}$$

This formula can be shortened to:

$$z = \frac{X - \overline{X}}{s}$$

Bob Fernandez got a score of 35.4. If the mean score was 25 and the standard deviation was 8, find his z-score.

$$z = \frac{X - \overline{X}}{s} = \frac{35.4 - 25}{8} = \frac{10.4}{8} = 1.30$$

Using his z-score, we can use Table A to find Bob's percentile rank. Column A at the extreme left lists the z-scores. Keep moving down column A until you find a z-score of 1.30. Just to the right of 1.30, in column B (area between mean and z) you'll find .4032. So .4032, or 40.32%, of the area lies between his score and the mean. Since 50% of the area also falls below the mean in a symmetrical distribution, we may conclude that 90.32% (50% + 40.32%) of all the area falls below a score of 35.4. And so, Bob Fernandez has a percentile rank of 90.32.

I didn't ask you to draw a sketch with that problem, but for a while it would be a good idea to do so. So, for the rest of the problems in this chapter, let's start out by drawing a normal curve so we can get a clear picture of what we're trying to find.

Here's a similar problem. If Kim Song Rhee weighed 24.7 pounds at one year of age, and the mean weight for one–year–olds is 21.2 pounds, find his percentile rank if the standard deviation is 4 pounds.

Figure 8.9

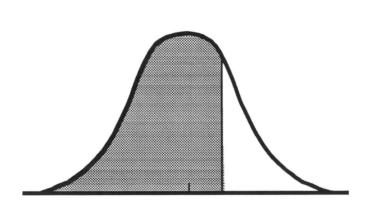

$$\overline{X} = 21.2 \quad X = 24.7$$

$$z = \frac{X - \overline{X}}{s} = \frac{24.7 - 21.2}{4} = \frac{3.5}{4} = .875$$

Since Table A lists z-scores to just two decimal places, what should we do? We could round up to .88, especially if we're in a hurry or if we need just an approximate answer. But I'm kind of a stickler for accuracy, so I'd prefer that we did a little interpolation. You may remember how to do this from Chapter 2.

In column B of Table A, we see that a z-score of 0.87 shows an area between the mean and z of .3078. And a z-score of 0.88 shows an area between the mean and z of .3106. So for a z-score of 0.875, we'll take an average of these two, which is .3092 (add .3078 and .3106 and divide by 2).

Now we can find our percentile rank. At the mean the percentile rank is 50%. Because Kim is 30.92% beyond the mean, he has a percentile rank of 80.92 (.50 + .30.92).

Are you getting the idea? First draw a sketch, then find the z-score, and finally, find the percentile rank. Moving right along, Jane O'Connor ran 247.1 miles in a six–day road race. If the mean distance run was 266.3 miles and the standard deviation was 29.6, what was her percentile rank?

Figure 8.10

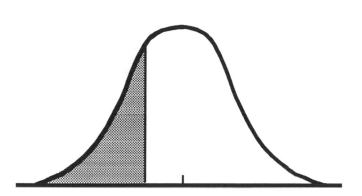

$$X = 247.1 \quad \overline{X} = 266.3$$

$$z = \frac{X - \overline{X}}{s} = \frac{247.1 - 266.3}{29.6} = \frac{-19.2}{29.6} = -.65$$

We can round off this z-score to –0.65. This is a negative score, so we're below the mean and the 50th percentile. In column B, the area between the mean and z is .2422. To find the percentile rank of 247.1 we'd have to subtract .2422 from .5000 to get .2578. But there's a much easier way to find this number. Just look at the number in column C—it's .2578. The heading of column C is the area beyond z.

Take a look at Figure 8.10. The area beyond z actually gives us our percentile rank, which happens to be 25.78. In other words, Jane O'Connor ran farther than 25.78% of the runners. On the other hand, of course, 74.22% of the runners ran further than she did. I don't know about you, but I wish I could run 247 miles in six days—or even six weeks.

Now find the percentile rank of Judy Cohen, who earns $42,000 at a company with an average salary of $59,000, and a standard deviation of $10,000.

Figure 8.11

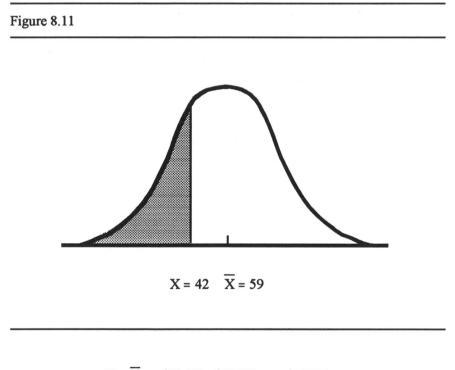

$$X = 42 \quad \overline{X} = 59$$

$$z = \frac{X - \overline{X}}{s} = \frac{\$42,000 - \$59,000}{\$10,000} = \frac{-\$17,000}{\$10,000} = -1.70$$

When the z-score is –1.7, the area beyond z in column C is .0446. So the percentile rank of Judy Cohen's salary is 4.46. Although she earns $42,000, over 95% of her colleagues earn more.

If Charles Yah Lin Trie scores 136 on a standard IQ test that has a mean of 100 and a standard deviation of 16, then what percent of people who took the test had higher IQ scores?

Figure 8.12

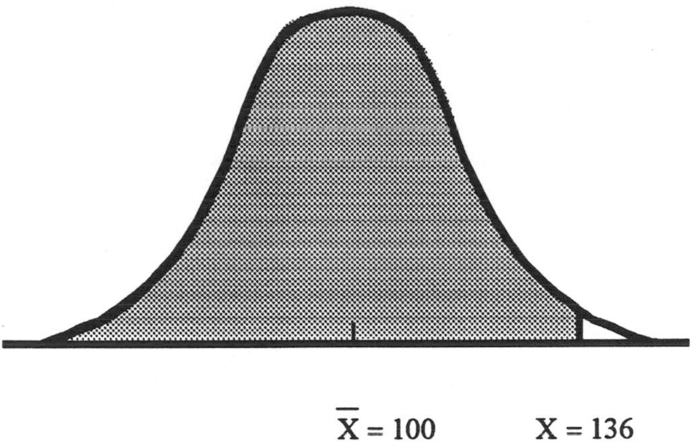

$$\overline{X} = 100 \qquad X = 136$$

$$z = \frac{X - \overline{X}}{s} = \frac{136 - 100}{16} = \frac{36}{16} = \frac{9}{4} = 2.25$$

We find from column C of Table A an area of .0122 beyond z. So 1.22% of the people taking the test had higher IQ scores than Mr. Trie.

Christine Laganis was out sick 2 days during the last year. At her company the average number of days out sick per employee is 6.78, and the standard deviation was 4.13. What is the percentile rank of Ms. Laganis's sick days relative to other employees? (Assume that those with the lowest number of sick days taken have the highest percentile rankings.)

Figure 8.13

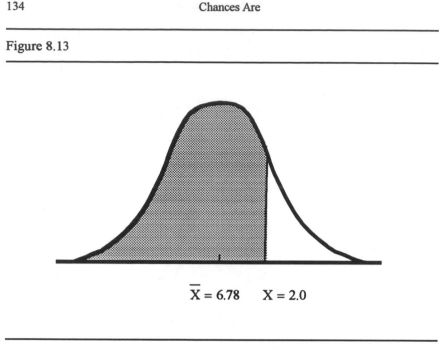

$$\overline{X} = 6.78 \qquad X = 2.0$$

$$z = \frac{X - \overline{X}}{s} = \frac{2 - 6.78}{4.13} = \frac{-4.78}{4.13} = -1.16$$

A z-score of –1.16 gives us an area of .1230 beyond z from column C of Table A. So Ms. Laganis's percentile rank was 12.30. In other words, she had a better attendance record than 87.7% of her fellow employees.

Patrick Ewing, the center for the New York Knicks, is 7'1" (that is, seven feet and one inch). If the average height of National Basketball Association players is 6'7.5" inches and the standard deviation is 5.3 inches, what percentage of NBA players is taller than Ewing?

Figure 8.14

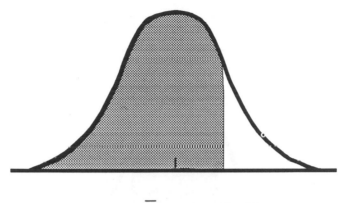

$$\overline{X} = 79.5 \qquad X = 85$$

$$z = \frac{X - \overline{X}}{s} = \frac{85 - 79.5}{5.3} = \frac{6.5}{5.3} = -1.23$$

Our first step is to convert our height measurements into inches. Seven foot one is 85 inches. Six foot seven and a half is 79.5 inches.

A z-score of 1.23 gives us an area beyond z of .1093. So 10.93% of all NBA players are taller than Patrick Ewing.

Finding Percent of Cases Falling between Two Scores

So far the problems we've been solving have been pretty straightforward. Draw a diagram, plug some numbers into the z-score formula, solve for the z-score, and look up the answer in Table A. Now we'll use the same tools to solve more complex problems.

We've set up a problem in Figure 8.15. What percent of cases fall between scores of 88 and 120 in a normal distribution that has a mean of 100 and a standard deviation of 16? This is really a three–part problem. First find the area between the scores of 88 and 100. Then find the area between the scores of 88 and 100. Then find the area between the scores of 120 and 100. Add the two areas and you've got your answer—the percent of cases that fall between scores of 88 and 120.

Figure 8.15

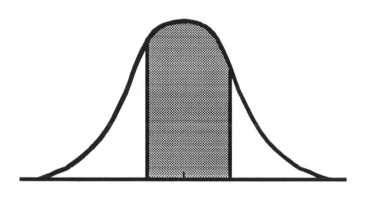

$$X = 88 \quad \overline{X} = 100 \quad X = 120$$

---

(1) $z = \dfrac{X - \overline{X}}{s} = \dfrac{120 - 100}{16} = \dfrac{20}{16} = 1.25$

Area between the mean and z = 1.25 = 39.44%

(2) $z = \dfrac{X - \overline{X}}{s} = \dfrac{88 - 100}{16} = \dfrac{-12}{16} = -0.75$

Area between the mean and z = –0.75 = 27.34%

(3)   39.44% + 27.34% = 66.78%

Next problem: Find the percent of cases between scores of 50 and 115 when the mean is 70 and the standard deviation is 25. Be sure to draw a sketch of the problem before you do any calculations.

Figure 8.16

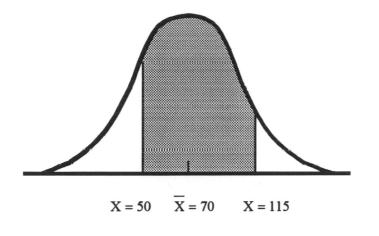

$$X = 50 \quad \overline{X} = 70 \quad X = 115$$

(1) $\quad z = \dfrac{X - \overline{X}}{s} = \dfrac{50 - 70}{25} = \dfrac{-20}{25} = -.80$

Area between the mean and z = –0.80 = 28.81%

(2) $\quad z = \dfrac{X - \overline{X}}{s} = \dfrac{115 - 70}{25} = \dfrac{45}{25} = 1.80$

Area between the mean and z = 1.80 = 46.41%

(3) 28.81% + 46.41% = 75.22%

Problem: Find the percent of cases between scores of 10 and 44 when the mean is 21 and the standard deviation is 9.

Figure 8.17

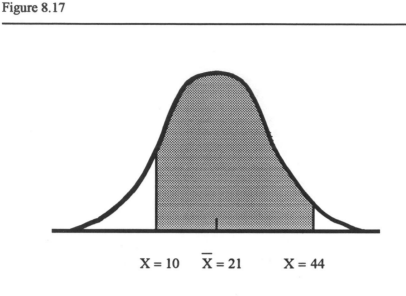

$$X = 10 \quad \overline{X} = 21 \qquad X = 44$$

(1) $z = \dfrac{X - \overline{X}}{s} = \dfrac{10 - 21}{9} = \dfrac{-11}{9} = -1.22$

      Area between the mean and z = –1.22 = 38.88%

(2) $z = \dfrac{X - \overline{X}}{s} = \dfrac{44 - 21}{9} = \dfrac{23}{9} = 2.56$

      Area between the mean and z = 2.56 = 44.48%

(3)   38.88% + 49.48% = 88.36%

And now for something a little bit different. We're going to find the percentage of people who had very high scores and very low scores on an IQ test. This test had a mean of 100 and a standard deviation of 20. Find the percentage of people who had scores of more than 150 or less than 70. First draw your sketch and then do your calculations.

Figure 8.18

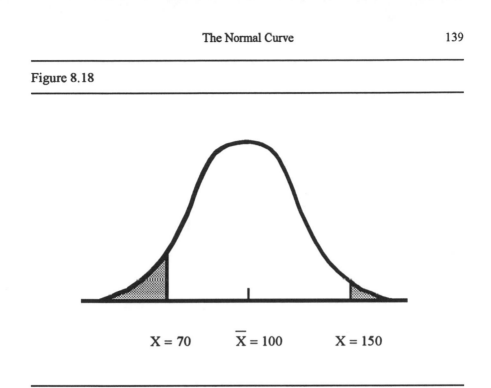

$$X = 70 \qquad \overline{X} = 100 \qquad X = 150$$

(1) $z = \dfrac{X - \overline{X}}{s} = \dfrac{150 - 100}{20} = \dfrac{50}{20} = 2.50$

Area beyond $z = 2.50 = .0062$.

(2) $z = \dfrac{X - \overline{X}}{s} = \dfrac{70 - 100}{20} = \dfrac{-30}{20} = -1.50$

Area beyond $z = -1.50 = .0668$

(3) $0.62\% + 6.68\% = 7.3\%$

Are you getting the hang of it? Here's another one to figure out. The average salary at the XYZ Corporation is $31,000. If the standard deviation is $9,500, what percentage of a normal distribution of employees earns more than $45,000 or less than $15,000?

Figure 8.19

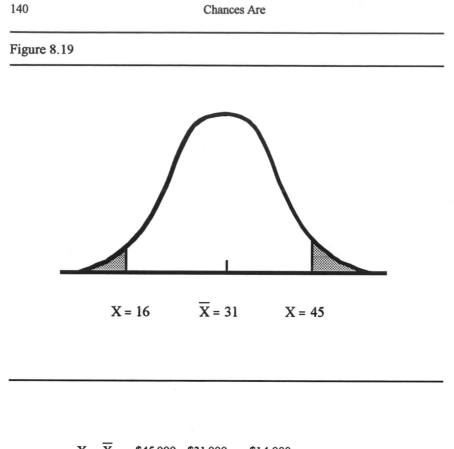

$$X = 16 \qquad \overline{X} = 31 \qquad X = 45$$

(1) $z = \dfrac{X - \overline{X}}{s} = \dfrac{\$45{,}000 - \$31{,}000}{\$9{,}500} = \dfrac{\$14{,}000}{\$9{,}500} = 1.47$

Area beyond z = 1.47 = .0708

(2) $z = \dfrac{X - \overline{X}}{s} = \dfrac{\$15{,}000 - \$31{,}000}{\$9{,}500} = \dfrac{-\$16{,}000}{\$9{,}500} = -1.68$

Area beyond z = –.0465

(3)   7.08% + 4.65% = 11.73%

Ready for still another type of problem? We've got a normal distribution of IQs, with a mean of 100 and a standard deviation of 18, and we want to find the percentage of IQs between scores of 125 and 150. See what you can do with this.

Figure 8.20

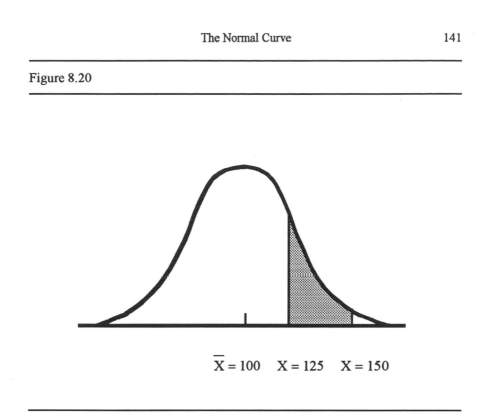

$$\overline{X} = 100 \quad X = 125 \quad X = 150$$

You can do this problem in three steps. First, find the percent of IQs between the mean and 150. Then find the percent of IQs between the mean and 125. Finally, subtract the percent between the mean and 125 from the percent between the mean and 150.

(1) $\quad z = \dfrac{X - \overline{X}}{s} = \dfrac{150 - 100}{18} = \dfrac{50}{18} = 2.78$

Area between the mean and z = 2.78 = 49.73%

(2) $\quad z = \dfrac{X - \overline{X}}{s} = \dfrac{125 - 100}{18} = \dfrac{25}{18} = 1.39$

Area between the mean and z = 1.39 = 41.77%

(3) $\quad$ 49.73% – 41.77% = 7.96%

Here's a similar problem. The ideal weight for a high school football lineman is between 215 and 280. The coach needs to know what percentage of the boys in the freshman class are in that weight category. If the weights of these boys are normally distributed, with a mean of 156.8 and a standard deviation of 33.4, what percentage of the freshman boys weigh between 215 and 280?

Figure 8.21

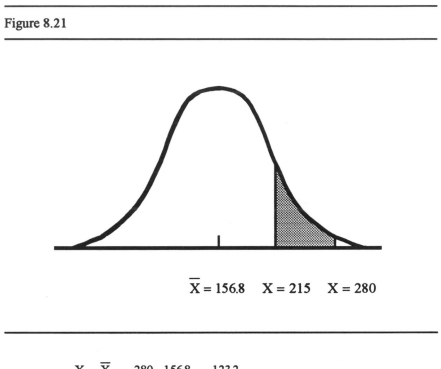

$$\overline{X} = 156.8 \quad X = 215 \quad X = 280$$

---

(1) $z = \dfrac{X - \overline{X}}{s} = \dfrac{280 - 156.8}{32.3} = \dfrac{123.2}{33.4} = 3.69$

Area between the mean and z = 3.69 = .4999

(2) $z = \dfrac{X - \overline{X}}{s} = \dfrac{215 - 156.8}{33.4} = \dfrac{58.2}{33.4} = -1.74$

Area between the mean and z = 1.44 = .4591

(3)   49.99% − 45.91% = 4.08%

And here's one more problem. The average height of third graders is 3'9.4". What percent of a normal distribution of third graders is between 3' tall and 3'6", if the standard deviation is 4.6"?

Figure 8.22

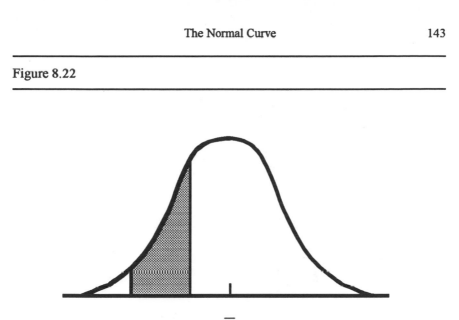

$$X = 36 \quad X = 42 \quad \overline{X} = 45.4$$

(1) $\quad z = \dfrac{X - \overline{X}}{s} = \dfrac{36 - 45.4}{4.6} = \dfrac{-9.4}{4.6} = -2.04$

Area between the mean and $z = -2.04 = .4793$

(2) $\quad z = \dfrac{X - \overline{X}}{s} = \dfrac{42 - 45.4}{4.6} = \dfrac{-3.4}{4.6} = -0.74$

Area between the mean and $z = -0.74 = .2704$

(3) $\quad 47.93\% - 27.04\% = 20.89\%$

Let's do one more. The average age of the residents of the Happy Day Rest Home is 86.3. If their ages are normally distributed with a standard deviation of 12.7, what percentage of the residents is between 65 and 75?

Figure 8.23

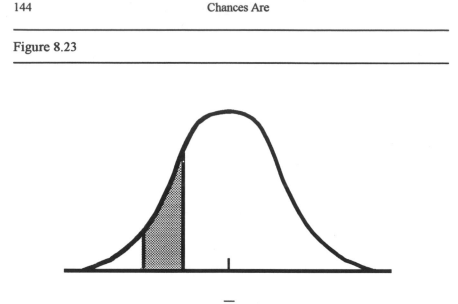

$$X = 65 \quad X = 75 \quad \overline{X} = 86.3$$

(1) $\quad z = \dfrac{X - \overline{X}}{s} = \dfrac{65 - 86.3}{12.7} = \dfrac{-21.3}{12.7} = -1.68$

Area between the mean and z = –1.68 = .4535.

(2) $\quad z = \dfrac{X - \overline{X}}{s} = \dfrac{75 - 86.3}{12.7} = \dfrac{-11.3}{12.7} = -0.89$

Area between the mean and z = –0.89 = .3133.

(3) $\quad$ 45.35% – 31.33% = 14.02%

As you've just seen, there are several different types of normal distribution problems you may be called upon to solve. Unfortunately no one will tell you exactly which type of problem you'll need to solve. In other words, in real life, our problems are mainly of the mix-and-match variety. So that's what we're going to throw at you in the chapter review. But don't worry; there's nothing there that you haven't already done.

Chapter Review

1.   Find the percentile rank of someone who weighs 140 pounds when the standard deviation is 28 and the mean is 104 pounds.

2.   Mayor Rudy Giuliani was mentioned favorably in the *New York Times* 297 times during his first term. If the average number of favorable mentions for a first–term mayor is 338 times and the standard deviation is 20.9, what is his percentile ranking among first–term mayors?

3.   Hector Perez ate 29 hot dogs in a contest. If the average number of hot dogs eaten was 17 and the standard deviation was 7.8, what percentage of the contestants ate more hot dogs than Mr. Perez?

4.   If the mean IQ in a normal distribution is 100, what percentage of scores would fall between IQs of 90 and 110 if the standard deviation were 16?

5.   If the average salary in a company is $52,500 and the standard deviation is $7,800, what percentage of employees earn between $40,000 and $50,000?

6.   Hiro Fukado is 4'3". If the average height in his fifth grade class is 4'6" and the standard deviation is 5.5", what is his percentile rank in the class?

7.   If the mean grade on an exam is 83 and the standard deviation is 9.4, what percentage of students got grades between 70 and 80?

8.   On an IQ test with a mean of 100 and a standard deviation of 18, find the percentage of people who had scores of more than 145 or less than 75.

Solutions:

1. $z = \dfrac{X - \overline{X}}{s} = \dfrac{140 - 104}{28} = \dfrac{36}{28} = 1.29 = .4015$

50% + 40.15% = 90.15 percentile

2. $z = \dfrac{X - \overline{X}}{s} = \dfrac{297 - 338}{20.9} = \dfrac{-41}{20.9} = -1.96 = .0250$

2.5 percentile

3. $z = \dfrac{X - \overline{X}}{s} = \dfrac{29 - 17}{7.8} = \dfrac{12}{7.8} = 1.54 = .0618$

= 6.18%

4. $z = \dfrac{X - \overline{X}}{s} = \dfrac{90 - 100}{16} = \dfrac{-10}{16} = -0.625$

−0.62 = .2324   − 0.63 = .2357

−0.625 = .23405 = 23.405% × 2 = 46.81%

5.   (1) $z = \dfrac{X - \overline{X}}{s} = \dfrac{\$40,000 - \$52,500}{\$7,800} = \dfrac{-\$12,500}{\$7,800} = -1.60 = .4452$

     (2) $z = \dfrac{X - \overline{X}}{s} = \dfrac{\$50,000 - \$52,500}{\$7,800} = \dfrac{-\$2,500}{\$7,800} = -0.32 = .1255$

     (3)   44.52% − 12.55% = 31.97%

6. $z = \dfrac{X - \overline{X}}{s} = \dfrac{51 - 54}{5.5} = \dfrac{-3}{5.5} = -0.55 = 29.12$ percentile

7.   (1) $z = \dfrac{X - \overline{X}}{s} = \dfrac{70 - 83}{9.4} = \dfrac{-13}{9.4} = -1.38 = .4162$

     (2) $z = \dfrac{X - \overline{X}}{s} = \dfrac{80 - 83}{9.4} = \dfrac{-3}{9.4} = -0.32 = .1255$

     (3)   41.62% − 12.55% = 29.07%

8.   (1) $z = \dfrac{X - \overline{X}}{s} = \dfrac{145 - 100}{18} = \dfrac{45}{18} = 2.50 = .0062$

     (2) $z = \dfrac{X - \overline{X}}{s} = \dfrac{75 - 100}{18} = \dfrac{-25}{18} = -1.39 = .0823$

     (3)   0.62% + 8.23% = 8.85%

Chapter 9

Probability

You will find much of probability theory very familiar. It can usually be reduced to questions like: What are the odds that an event will occur? Or what is the likelihood that such and such will happen?

For instance, if you flip a coin, what are the odds that it will come up heads? Obviously the odds are even—1 to 1, or 1:1. And what is the likelihood of drawing a diamond from a well-shuffled deck of playing cards? It's one out of four, or 1/4.

Incidentally, what are the odds of tossing a coin twice and coming up with two heads? Intuitively you may have come up with the answer one out of four, or 1/4, and you'd be right. This raises a very important question about events occurring in sequence: Are these events *independent*? Two events are said to be independent if the selection of one has no effect upon the probability of selecting the other event. In other words, if you flip a coin and it comes up heads, the outcome of your first flip has absolutely no bearing on the outcome of your second flip.

All we need to do now is define probability and we're ready to do some problems. Let's look at the probability of getting heads when you toss a coin. The probability of getting a head (event H) =

$$\frac{\text{number of outcomes favoring event H}}{\text{total number of events (those favoring H and those not favoring H)}}$$

Think of that probability as a proportion. How much *is* that particular proportion? It's 1/2, or .5. Indeed, the probability of an event is always a number between 0 and 1. If an event is certain to occur, what is its probability? Its probability is 1. And what is that event's probability if it is certain *not* to occur? Then its probability is 0.

Sampling with Replacement

Are you ready to do a statistical experiment? Don't worry, you won't need any test tubes or a Bunsen burner, and I promise to do all the work. Well, *most* of the work. First write out the numbers 0, 1, 2, and 3 on scraps on paper and place them in a box, a hat, or your pocket.

Now pull out one scrap and record its number on a sheet of paper. Replace the scrap of paper in the box, hat, or pocket and make sure the scraps are mixed. Now pull out a scrap and record the number on your sheet of paper.

That's all you're going to have to do in this experiment. I'll do all the work from here on. The first thing I'm going to do is record all the possible outcomes from our sample. All we'll be doing is pulling scraps of paper from the box. We'll pull one scrap, record the number, replace the scrap in the box, and then pull a second scrap and record the number. In Table 9.1 I've summarized all the possible outcomes of our experiment.

Table 9.1: Possible Outcomes of Pulling Scraps Numbered 0, 1, 2, and 3 from a Box at Random

---

Outcomes starting with 0	0,0	0,1	0,2	0,3
Outcomes starting with 1	1,0	1,1	1,2	1,3
Outcomes starting with 2	2,0	2,1	2,2	2,3
Outcomes starting with 3	3,0	3,1	3,2	3,3

There are 16 possible outcomes listed in Table 9.1. Now think about it. If we picked these scraps of paper out of a box, the probability of picking *any* pair would be equal to the probability of picking *any other* pair. For example, the probability of picking (1,3) would be exactly the same as the probability of picking (2,0). Or (3,2). Or (0,1).

Are you with me so far? OK. Then what *is* the probability of picking (1,3)?

The probability of picking (1,3) is 1/16, or one chance in 16. Where did we get *that* from? Well, there are 16 possible outcomes listed in Table 9.1. (If you don't trust me, just add up all the possible outcomes yourself.) Going back to our probability equation, the probability of an event H =

$$\frac{\text{number of outcomes favoring event H}}{\text{total number of events (those favoring H and those not favoring H)}}$$

So the probability of picking (1,3) is 1/16. The probability of picking, say, (2,0) is also 1/16. And the probability of picking, at random, *any* pair of numbers is 1/16.

Now we're going to take our probability analysis to a higher level. We're going to set up a frequency and probability distribution—an endeavor that is a lot simpler than it sounds.

All we're really going to do is rearrange the data from Table 9.1 into a frequency distribution showing the probabilities of drawing different sets of numbers. We're going to answer seven questions: What is the probability of drawing two numbers whose sum is 0, 1, 2, 3, 4, 5, or 6? Table 9.2 gives us the frequency distribution that we need to answer these questions.

Table 9.2: Frequency Distribution of Number Sets by Sums of Numbers in Set

Sum	Set	Frequency
0	0,0	1
1	0,1 1,0	2
2	0,2 1,1 2,0	3
3	0,3 1,2 2,1 3,0	4
4	1,3 2,2 3,1	3
5	2,3 3,2	2
6	3,3	1

Now we've got all the data we need to do our frequency and probability distribution, which will answer the seven questions: What is the probability of drawing two numbers whose sum is 0, 1, 2, 3, 4, 5, or 6? See if you can come up with the answers yourself. Then check out my answers in Table 9.3.

Table 9.3: Frequency and Probability Distribution of Number Sets by Sums of Numbers in Set

Sum	Frequency	Probability
0	1	.0625
1	2	.125
2	3	.1875
3	4	.25
4	3	.1875
5	2	.125
6	1	.0625

Using Table 9.3, what is the probability of drawing a set of numbers whose sum is 4? The probability is .1875, which translates to 18.75 chances in 100. And what is the probability of drawing a set of numbers whose sum is 5? The answer is .125, or 12.5%.

I want you to keep in mind that the probabilities must add up to 1.0. This is a good check on your answers. If your probabilities don't add up to exactly 1.0, then you've made a mistake that you'll need to go back to and correct. Of course, if the sum of your probabilities comes to slightly more or less than 1.0, then this discrepancy may be due to rounding and not to a mistake.

You'll notice I've converted these probability fractions into decimals. I needed to do that for a couple of reasons. First, probability is usually stated as a proportion of 1, and in statistics the conventional way of stating a proportion is a decimal. Hey, don't get mad at *me*. I don't make up the rules; I just follow them. The second reason we state probability as a decimal is that decimals are easier to use than fractions in drawing graphs.

And that's exactly what we're going to do next. Do you remember how to draw a histogram? Just check back in Chapter 1. I'd like you to draw a histogram of the probabilities of picking number sets with sums of 0, 1, 2, 3, 4, 5, and 6. See if your histogram looks like the one I've drawn in Figure 9.1.

Figure 9.1: Probability Histogram of Sums of Pairs of Numbers Based upon Selecting with Replacement from a Population of Four Numbers

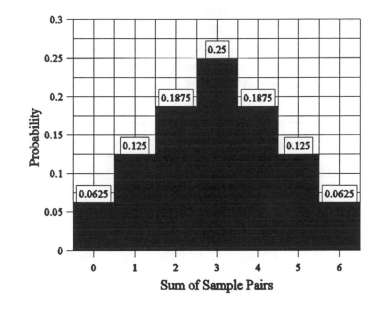

If you flipped a coin twice, what are the possible outcomes? Make a chart, starting with two heads, or HH.

Here's my chart:

HH
HT
TT
TH

What's the probability of getting no heads, one head, and two heads?

The probability of getting no heads is 1/4, or .25. The probability of getting one head is 2/4, or .5. And the probability of getting two heads is 1/4, or .25.

Now see if you can figure out all the outcomes of flipping a coin three times. Make a chart and then find the probability of getting no heads, one head, two heads, and three heads.

Here's my chart:

HHH
HHT
HTH
HTT
TTT
TTH
THT
THH

There are a total of eight possible outcomes. Using this data, I can do my frequency and probability distribution table (see Table 9.4).

Table 9.4: Frequency and Probability of Getting Heads on Three Coin Tosses

Heads	Frequency	Probability
0	1	.125
1	3	.375
2	3	.375
3	1	.125

We can see that the probability of getting no heads (or three tails in a row) is .125; the probability of getting one head is .375; the probability of getting two heads is .375; and the probability of getting three heads is .125.

The Addition Rule

Solving these probability problems is not only a whole lot of fun, but it's actually leading up to something. It's leading up to the addition rule. But don't get too excited, because we're not quite there yet.

Let's go back to our last problem. If we flip a coin three times, what is the probability of getting two or more heads?

The answer is .5, or 50%. How did we get it? All we needed to do was to look it up in Table 9.4. We added the probability of getting two heads, .375, and the probability of getting three heads, .125.

Now we'll go back to our previous problem in which we drew a pair of numbers from a hat. Using Table 9.3, find the probability of drawing a pair of numbers whose sum is 3 or less.

The answer is .625, or 62.5%. We got our answer by adding the probabilities of drawing a pair of numbers whose sum was 0 (.0625), 1 (.125), 2 (.1875), and 3 (.25).

In these last two problems we added probabilities and we saw that it was good. It is OK to add separate probabilities as long as the events are *mutually exclusive*.

Are you ready for the addition rule? All right then, here it comes: If X and Y are mutually exclusive events, the probability of obtaining either of them is equal to the probability of X plus the probability of Y. And as we've already seen, this formula can be extended to include any number of mutually exclusive events.

The Multiplication Rule

Moving right along, what is the likelihood of tossing a coin and getting a tail. It's 1/2, or .5. What is the likelihood of tossing a coin and getting two tails in a row? Figure it out. Do a chart.

The probability of two tails in a row would be 1/4, or .25:

> TT
> TH
> HT
> HH

Now, what is the probability of getting three tails in a row? We did that one before, so go back a couple of pages and check my chart.

The probability of getting three tails in a row is 1/8, or .125.

Now here's the multiplication rule: The probability of the simultaneous or successive occurrence of two events is the product of the separate probabilities of each event. We can expand that to read: The probability of the simultaneous or successive occurrence of two or more events is the product of the separate probabilities of each event.

The occurrence of event X is not dependent upon the occurrence of event Y, or event Z. Similarly, the occurrence of event Y is not dependent upon the occurrence of event X or event Z, and the occurrence of event Z is not dependent upon the occurrence of event X or event Y. These events are *independent*.

Let's go back to finding the probability of getting three tails in a row. We know the probability of flipping a coin and having it come up tails is 1/2, or .5. And we also know the probability of getting two tails in a row is 1/4, or .25. Now if we happen to multiply 1/2 by 1/2, we get 1/4, or, for that matter, if we multiply .5 by .5, we get .25.

What is the probability of getting three tails in a row? Just multiply $1/2 \times 1/2 \times 1/2 = 1/8$. Or multiply $.5 \times .5 \times .5 = .125$.

So the multiplication rule comes in pretty handy. What is the probability of getting five tails in a row? Work it out.

Solution: $1/2 \times 1/2 \times 1/2 \times 1/2 \times 1/2 = 1/32$. Or $.5 \times .5 \times .5 \times .5 \times .5$ = .03125.

Sampling without Replacement

So far we've been doing sampling with replacement. Now we're going to sample without replacement, which changes the odds.

Get out that well-shuffled deck of playing cards. What are the chances of picking a queen?

There are 4 queens in a deck of 52 cards, so your chances of picking a queen are 4/52, or 1/13, or .076923. If you placed the card back in the deck and reshuffled the cards, your chances of picking a queen would remain 1/13, or .076923. But what if you did not replace the card? What would your chances be of drawing a queen?

They would be only 3/51, or .0588235. OK, what is your probability of drawing two queens in a row if there's no replacement? See if you can figure out the answer (and you may definitely use a calculator).

I got .0045248. If you got a slightly different answer, that's fine, because the difference between our answers was due to rounding.

Now just for the fun of it, what is the probability of your drawing two queens if there is replacement?

The answer is .0059171. So your chances of drawing two queens are somewhat greater with replacement than without replacement.

Let's try another card trick. What are the chances of drawing three spades in a row from a deck of well-shuffled playing cards if there is no replacement?

Here's the solution: $13/52 \times 12/51 \times 11/50 = 1716/132,600 = .0129411$. If your answer is slightly different from mine, the difference is probably due to rounding and is nothing to lose any sleep over.

We've alluded here and there to sampling. Picking a scrap of paper with a number on it from a box is sampling. So is drawing cards from a deck. And we've done a whole dog and pony show to illustrate the importance of the difference between sampling with and without replacement.

So how important *is* the distinction between sampling with and without replacement? That depends on the size of the population from which we are drawing our sample. If our population is just four scraps of paper with numbers written on them, then the issue of replacement is extremely important. Whether we replace a card in a 52-card deck is very important to the size of our probability. But when you're selecting from a very large population of scores, the difference between sampling *with* and *without* replacement becomes trivial. For example, if we took one thousand decks of playing cards, shuffled them, and merged the decks into one 52,000-card deck, how much would it affect our probabilities of drawing two or three queens in a row if we replaced the cards we drew? It would make only a tiny difference—so tiny, in fact, that it would be trivial.

Double Counting

What is the probability of drawing either a heart or a king from a deck of cards? Pretty tricky, eh? Let's start with the heart. The chances of drawing a heart are 13/52. And the chances of drawing a king are 4/52. But there's a 1/52 chance you'll draw the king of hearts. We need to eliminate this joint occurrence:

$$13/52 + 4/52 - 1/52 = 16/52 = 4/13 = .30769$$

Here's another problem. One quarter of the soldiers at an army base are black and one tenth of the soldiers at that base are female. What are the chances of picking, at random, a soldier who is either black or female?

Solution: We start with a 1/4 probability of selecting a black soldier and a 1/10 probability of selecting a female soldier. What is the probability of selecting a black female soldier? It's $1/4 \times 1/10 = 1/40$. So the probability of selecting a black soldier or a female soldier would be $1/4 + 1/10 - 1/40 = 10/40 + 4/40 - 1/40 = 13/40 = .325$.

That was so much fun, let's try one more problem. A school with 1,000 students has a football team with 50 members and a baseball team with 25 members. There are two people who play both baseball and football. What is the probability of selecting someone who plays either baseball or football?

Solution: The probability of picking a football player is 50/1000, and the probability of picking a baseball player is 25/1000. But the probability of picking someone who plays baseball or football is $50/1000 + 25/1000 - 2/1000 = 73/1000 = .073$.

Probability and the Normal Curve

The standard normal curve lends itself very well to probability problems. Indeed, the problems you'll be solving here are very similar to those you solved in the previous chapter.

Here's our first problem. What is the probability of selecting a person at random from the general population who has an IQ of at least 120? Assume that $\bar{X} = 100$ and the standard deviation is 14. Remember to sketch a normal curve before you do any calculations.

Figure 9.2

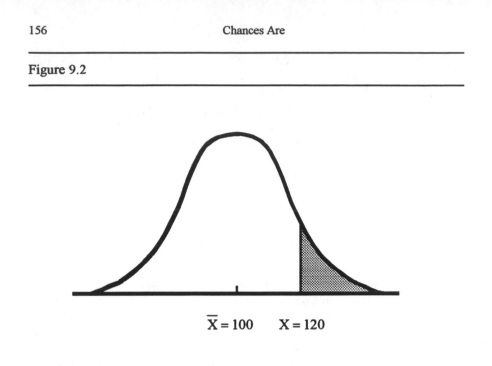

$$\overline{X} = 100 \qquad X = 120$$

$$z = \frac{X - \overline{X}}{s} = \frac{120 - 100}{14} = \frac{20}{14} = 1.43 = .0764$$

Using column C of Table A we find that .0764 of the area lies at or beyond a z-score of 1.43. Therefore the probability of selecting, at random, a score of at least 120 is .0764.

Using the same mean and standard deviation, what is the probability of selecting, at random, a person with an IQ score of at least 90?

Figure 9.3:

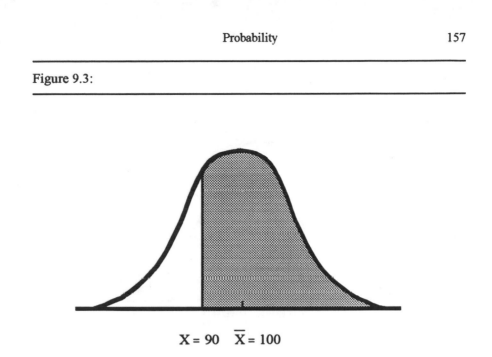

$$X = 90 \quad \overline{X} = 100$$

(1) $z = \dfrac{X - \overline{X}}{s} = \dfrac{90 - 100}{14} = \dfrac{-10}{14} = -0.71 = .2580$

(2) The area to the right of the mean is .5000.

(3) The probability of selecting, at random, a score of at least 90 is .7580.

     Next problem: Using the same mean and standard deviation, what is the probability of selecting, at random, three people with IQs equaling or exceeding 125? To solve this problem you'll need to use the multiplication rule.

Figure 9.4

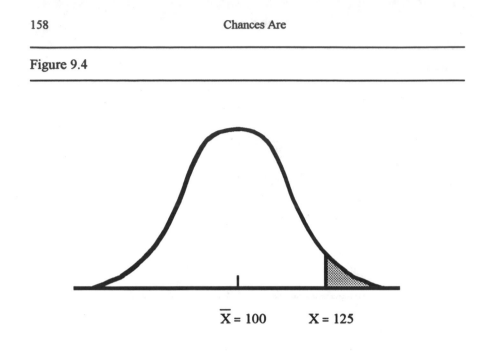

$$\overline{X} = 100 \qquad X = 125$$

$$z = \frac{X - \overline{X}}{s} = \frac{125 - 100}{14} = \frac{25}{14} = 1.79 = .0367$$

$.0367 \times .0367 \times .0367 = .0000494$

This translates to slightly less than five chances in 100,000.

One more problem: Using the same mean and standard deviation, what is the probability of selecting, at random, four people with IQs equal to or less than 105?

(1) $\quad z = \dfrac{X - \overline{X}}{s} = \dfrac{105 - 100}{14} = \dfrac{5}{14} = 0.36 = .1406$

(2) $\quad .1406 + .5000 = .6406$

(3) $\quad .6406 \times .6406 \times .6406 \times .6406 = .1684$

You probably noticed that I didn't bother to sketch this problem. If you're not quite sure what you're doing, it's very helpful to draw a picture of the problem, but as you gain experience you won't need to do this.

One- and Two-Tailed Probability Val'

So far our normal curve probability problems all ca
probability values that were equal to or higher than a particular scoι,
were equal to or lower than a particular score. For instance, we found in tιι.
problem the probability of picking someone with an IQ of 125 or higher. And in
the last problem we found the probability of selecting four people with IQs equal
to or less than 105.

Now, instead of looking for the probability of selecting scores that are either
higher than the mean or *lower* than the mean, we're going to look for the
probability of selecting scores that are both higher *and* lower than the mean. Don't
worry, finding these scores is not nearly as complicated as it sounds.

Let's stick with our IQ standard normal curve with a mean of 100 and a
standard deviation of 14.

Problem: What is the probability of selecting a person who has an IQ equal
to or more than 140, or equal to or less than 60? Sketch a picture and then solve
the problem. Hint: Find the probability of selecting a person with an IQ of 140 or
more and then double your answer.

Figure 9.5

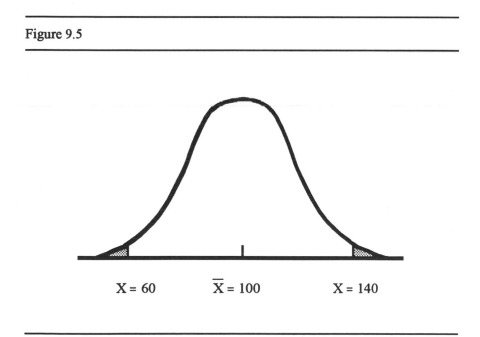

$X = 60$ $\overline{X} = 100$ $X = 140$

---

(1) $\quad z = \dfrac{X - \overline{X}}{s} = \dfrac{140 - 100}{14} = \dfrac{40}{14} = 2.86 = .0021$

(2) $\quad .0021 \times 2 = .0042$

Since selecting someone with an IQ equal to or less than 60 is as likely as selecting someone with an IQ equal to or more than 140, we just need to find the probability shown in one tail of Figure 9.5 and double it.

Next problem: What is the probability of selecting a person who has an IQ equal to or more than 115, or equal to or less than 85?

Figure 9.6

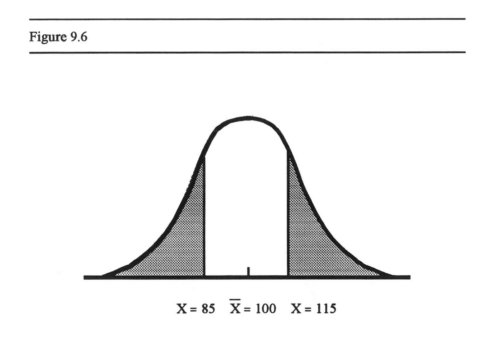

$$X = 85 \quad \overline{X} = 100 \quad X = 115$$

---

(1) $\quad z = \dfrac{X - \overline{X}}{s} = \dfrac{115 - 100}{14} = \dfrac{15}{14} = 1.07 = .1423$

(2) $\quad .1423 \times 2 = .2846$

Here's one that's somewhat more difficult. What is the probability of selecting three people who have IQs equal to or more than 105, or equal to or less than 95?

(1) $z = \dfrac{X - \overline{X}}{s} = \dfrac{105 - 100}{14} = \dfrac{5}{14} = 0.36 = .3594$

(2) $.3594 \times .3594 \times .3594 = .046423$

(3) $.046423 \times 2 = .0928$

Let's go over the solution step by step. (1) We find the probability of selecting one person with an IQ equal to or higher than 105. That probability is .3594. (2) We find that the probability of selecting three people with IQs equal to or higher than 105 is .046423. (3) We double that probability to find the probability of selecting people with IQs either equal to or higher than 105, or equal to or lower than 95.

Here's one last problem. What is the probability of selecting four people who have IQs equal to or higher than 108, or equal to or lower than 92?

(1) $z = \dfrac{X - \overline{X}}{s} = \dfrac{108 - 100}{14} = \dfrac{8}{14} = 0.57 = .2843$

(2) $.2843 \times .2843 \times .2843 \times .2843 = .0065329$

(3) $.0065329 \times 2 = .0131$

Chapter Review

1.   If you roll a set of dice, what are your chances of getting (a) 9; (b) 2; or (c) a 7 or an 11?

2.   Write the numbers 1, 2, 3, 4, and 5 on five scraps of paper and place them in a box. If you select one scrap from the box, record its number, put it back in the box, and then reach in and pick a scrap and record that number: (a) What are the chances that the sum of the two numbers you selected will be 4? (b) What are the chances that the sum of the two numbers you selected will be 10?

3.   If you flip a coin six times in a row, what is the probability of getting six heads?

4.   Using the data from problem 2, what is the probability of drawing a pair of numbers whose sum is 5 or less?

5.   Using the data from problem 2, what is the probability of drawing a pair of numbers whose sum is 7 or more?

6.   Using a well-shuffled deck of playing cards, what is the probability of drawing two clubs in a row (a) if there is replacement? (b) if there is no replacement?

7.   Using a well-shuffled deck of playing cards, what is the probability of drawing three picture cards in a row (a) if there is replacement? (b) if there is no replacement?

8.   What is the probability of drawing either a diamond or an ace from a deck of playing cards?

9.   At San Antonio College 30% of the students attend part-time and 20% of the students are business majors. What is the probability of selecting at random either a part-time student or a business major?

10.   There is a standard normal distribution of people with a mean of 114.8 pounds and a standard deviation of 26.3 pounds. What is the probability of selecting a person at random who weighs at least 150 pounds?

11.   Using the data from problem 10, what is the probability of selecting a person at random who weighs 140 pounds or less?

12.   Using the data from problem 10, what is the probability of selecting three people at random with weights equaling or exceeding 100 pounds?

13.   Assume a standard IQ normal curve with a mean of 100 and a standard deviation of 18. What is the probability of selecting a person who has an IQ equal to or more than 125, or equal to or less than 75?

14.   Using the data from problem 13, what is the probability of selecting four people who have IQs equal to or more than 105, or equal to or less than 95?

Solutions to Chapter Review

1.   (1,1)   (1,2)   (1,3)   (1,4)   (1,5)   (1,6)
      (2,1)   (2,2)   (2,3)   (2,4)   (2,5)   (2,6)
      (3,1)   (3,2)   (3,3)   (3,4)   (3,5)   (3,6)
      (4,1)   (4,2)   (4,3)   (4,4)   (4,5)   (4,6)
      (5,1)   (5,2)   (5,3)   (5,4)   (5,5)   (5,6)
      (6,1)   (6,2)   (6,3)   (6,4)   (6,5)   (6,6)

      (a)   (3,6) (4,5) (5,4) (6,3) = 4

$$\frac{4}{36} = \frac{1}{9} = .1111$$

      (b)   (1,1) = 1 $\frac{1}{36}$ = .0278

      (c)   (1,6) (2,5) (3,4) (4,3) (5,2) (6,1) = 6

                 (5,6) (6,5) = 2

$$\frac{8}{36} = \frac{2}{9} = .2222$$

2.   (a)   (1,3) (2,2) (3,1) = 3

$$\frac{3}{25} = .12$$

       (b)   (5,5) = 1

$$\frac{1}{25} = .04$$

3. $\frac{1}{2} \times \frac{1}{2} \times \frac{1}{2} \times \frac{1}{2} \times \frac{1}{2} \times \frac{1}{2} = \frac{1}{64}$

4.   (1,1) (1,2) (1,3) (1,4)

                 (2,1) (2,2) (2,3)

                 (3,1) (3,2)

                 (4,1)

$$\frac{10}{25} = .4$$

5.   (2,5)

                 (3,4) (3,5)

                 (4,3) (4,4) (4,5)

                 (5,2) (5,3) (5,4) (5,5)

$$\frac{10}{25} = .4$$

6.   (a) $\frac{1}{4} \times \frac{1}{4} \times \frac{1}{16} = .0625$

       (b) $\frac{1}{4} \times \frac{12}{51} = \frac{3}{51} = .0588$

7.   (a) $\frac{12}{52} \times \frac{12}{52} \times \frac{12}{52} = \frac{3}{13} \times \frac{3}{13} \times \frac{3}{13} = \frac{27}{2197} = .0123$

       (b) $\frac{12}{52} \times \frac{11}{51} \times \frac{10}{50} = \frac{1320}{132,600} = \frac{132}{13,260} = .0095$

8. $\dfrac{13}{52} + \dfrac{4}{52} - \dfrac{1}{52} = \dfrac{16}{52} = .3077$

9. $.3 + .2 - .06 = .44$

10. $z = \dfrac{X - \overline{X}}{s} = \dfrac{150 - 114.8}{26.3} = \dfrac{35.2}{26.3} = 1.34 = .0901$

11. (1) $z = \dfrac{X - \overline{X}}{s} = \dfrac{140 - 114.8}{26.8} = -\dfrac{25.2}{26.3} = 0.96 = .3315$

    (2) $.3315 + .5000 = .8315$

12. (1) $z = \dfrac{X - \overline{X}}{s} = \dfrac{100 - 114.8}{26.8} = -\dfrac{14.8}{26.8} = -0.55 = .2088$

    (2) $.2088 + .5000 = .7088$

    (3) $.7088 \times .7088 \times .7088 = .3561$

13. (1) $z = \dfrac{X - \overline{X}}{s} = \dfrac{125 - 100}{18} = \dfrac{25}{18} = 1.39 = .0823$

    (2) $.0823 \times 2 = .1646$

14. (1) $z = \dfrac{X - \overline{X}}{s} = \dfrac{105 - 100}{18} = \dfrac{5}{18} = 0.28 = .3897$

    (2) $.3897 \times 2 = .7794$

    (3) $.7794 \times .7794 \times .7794 \times .7794 = .3690$

Chapter 10

Correlation

Have you ever noticed that the taller someone is, the more that person will weigh? Of course we may know some short people who are very heavy, or some tall folks who are all skin and bones, but clearly there is a strong relationship between height and weight. Correlation is a statistical measurement of this relationship.

Take a look at the scatter diagram in Figure 10.1a. The correlation coefficient, which we'll introduce formally in a couple of pages, can be as high as 1.0 for a positive correlation, or as low as −1.0 for a negative correlation. So what would you estimate the correlation coefficient to be in the scatter diagram shown in Figure 10.1a?

Figure 10.1a: Hypothetical Sampling of the Heights and Weights of 260 Men, Aged 21–59

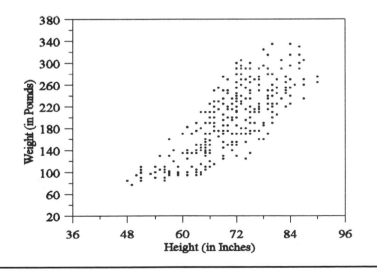

I'd give it about a 0.7, or maybe even an 0.8. But it's definitely not a perfect 1.0. Moving right along, what would you estimate the correlation coefficient to be in Figure 10.1b?

Figure 10.1b: Hypothetical Sampling of Grade Point Average and Hours Per Day Spent Watching TV of 250 College Students

---

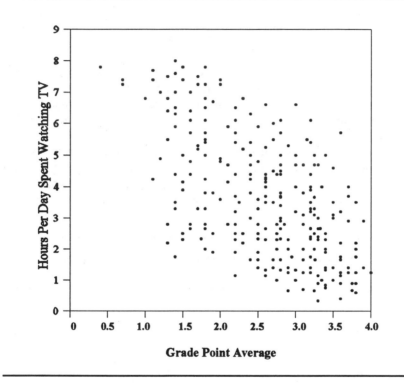

It's definitely negative. But it's not as pronounced as the one in Figure 10.1a. I'd give somewhere around a –0.5. Probably different observers looking at this scatter diagram would come up with a variety of answers. OK, now give me your best estimate of the correlation coefficient for the scatter diagram in Figure 10.1c.

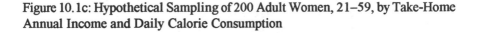

Figure 10.1c: Hypothetical Sampling of 200 Adult Women, 21–59, by Take-Home Annual Income and Daily Calorie Consumption

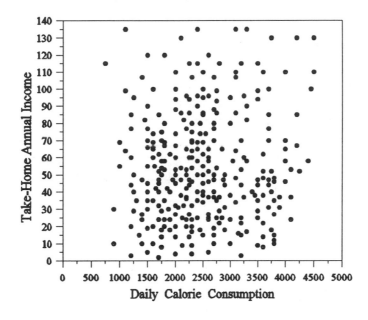

I can't see *any* correlation in Figure 10.1c between take-home annual income and daily calorie consumption. So I'd estimate the correlation to be 0. Don't worry—if you saw a slightly negative correlation of, say, –0.05 or a slightly positive correlation of, say, 0.1, then you're definitely in the ballpark. When we do visual correlation estimates, there rarely *is* just one correct answer.

Positive, Negative, and Perfect Correlations

The scatter diagram in Figure 10.1a is an example of a positive correlation. It shows that as a person's measured height rises, his weight tends to rise as well. And the scatter diagram in Figure 10.1b shows a negative correlation. As a college student's measured TV watching rises, her grade point average tends to decline.

A perfectly positive relationship has a correlation coefficient of +1.0, while a perfectly negative relationship has one of –1.0. And a correlation coefficient of 0 indicates that there is no relationship between two variables.

What would the scatter diagram of a perfectly positive relationship look like? It would look like a straight line moving from the lower left to the upper right, as shown in Figure 10.1d. And what would a perfectly negative relationship look like? It would look like a straight line moving from the upper left to the lower right, as shown in Figure 10.1e.

Figure 10.1d: Perfectly Positive Correlation

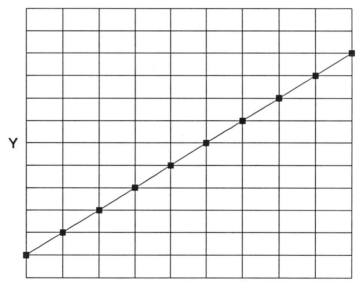

Figure 10.1e: Perfectly Negative Correlation

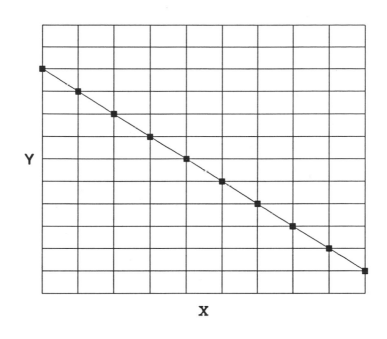

Y

X

A weak positive correlation would come in at 0.1, 0.2, or possibly 0.3. And a strong positive correlation would be anything above 0.6 or so. Similarly, a strong negative relationship would be –0.9, –0.8, or –0.7. And a weak negative relationship would be at –0.3, –0.2, or –0.1.

Calculating the Coefficient of Correlation

I'd like you to study the numbers in Table 10.1 and see if you can estimate the coefficient of correlation. Write it down and we'll see how close you came. Now please find the mean of the X's and the mean of the Y's.

Table 10.1

X	Y
2	9
4	10
5	12
8	15
11	19

$\overline{X}$ is 6 and $\overline{Y}$ is 13. Now we need to fill in three more columns in Table 10.2: X^2, Y^2, and XY. After you've done that, find $\sum X^2$, $\sum Y^2$, and $\sum XY$ by adding these three columns. Then check your work with mine in Table 10.3.

Table 10.2

X	X^2	Y	Y^2	XY
2		9		
4		10		
5		12		
8		15		
11		19		
30		65		
$\overline{X} = 6$		$\overline{Y} = 13$		

Table 10.3

X	X²	Y	Y²	XY
2	4	9	81	18
4	16	10	100	40
5	25	12	144	60
8	64	15	225	120
11	121	19	361	209
30	330	65	911	447

$\overline{X} = 6$ $\qquad\qquad$ $\overline{Y} = 13$

Do you remember what we were trying to find? We were trying to find the coefficient of correlation, which is designated by the letter r. All we need to do is plug a bunch of numbers into a formula and solve for r. But first, let's go over each of the terms. You already know ΣXY, and you found ΣX and ΣY when you calculated $\overline{X}$ and $\overline{Y}$. How much is n? It's 5, since we started with five pairs of X's and Y's back in Table 10.1. You already have ΣX^2, but you still need to find $(\Sigma X)^2$. Similarly you already know ΣY^2, but you still need to find $(\Sigma Y)^2$. So you've really got your work cut out for you. After you've solved for r, check my work below.

$$r = \frac{\Sigma XY - \dfrac{(\Sigma X)(\Sigma Y)}{n}}{\sqrt{\left[\Sigma X^2 - \dfrac{(\Sigma X)^2}{n}\right]\left[\Sigma Y^2 - \dfrac{(\Sigma Y)^2}{n}\right]}}$$

Solution:

$$r = \frac{447 - \dfrac{(30)(65)}{5}}{\sqrt{\left[330 - \dfrac{(30)^2}{5}\right]\left[911 - \dfrac{(65)^2}{5}\right]}}$$

$$r = \frac{447 - \dfrac{1950}{5}}{\sqrt{\left[330 - \dfrac{900}{5}\right]\left[911 - \dfrac{4225}{5}\right]}}$$

$$r = \frac{447 - 390}{\sqrt{[330 - 180]\,[911 - 845]}}$$

$$r = \frac{57}{\sqrt{(150)\,(66)}}$$

$$r = \frac{57}{\sqrt{9900}}$$

$$r = \frac{57}{99.4987}$$

$$r = 0.57$$

Our coefficient of correlation came to 0.57. Generally we round to two decimal places. So how close was your estimate? If you guessed at about 0.6 you were quite close. Before you work out one of these problems, always estimate the r. That way, if your answer is way off, you'll want to check your work. And if you were expecting a negative correlation and your r came out positive, then you know something is wrong. And of course, if your r does not fall within the range of −1.0 to +1.0, then you definitely need to rework the problem.

Here's another problem for you to work out in Table 10.4. After you've finished, check my solution.

Table 10.4

X	X²	Y	Y²	XY
3		15		
5		12		
8		10		
9		7		
11		8		
13		8		

Solution:

Table 10.5

X	X²	Y	Y²	XY
3	9	15	225	45
5	25	12	144	60
8	64	10	100	80
9	81	7	49	63
11	121	8	64	88
12	144	8	64	96
48	444	60	646	432

$\overline{X} = 8$ $\overline{Y} = 10$

$$r = \frac{\sum XY - \frac{(\sum X)(\sum Y)}{n}}{\sqrt{\left[\sum X^2 - \frac{(\sum X)^2}{n}\right]\left[\sum Y^2 - \frac{(\sum Y)^2}{n}\right]}}$$

$$r = \frac{432 - \frac{(48)(60)}{6}}{\sqrt{\left[444 - \frac{(48)^2}{6}\right]\left[646 - \frac{(60)^2}{6}\right]}}$$

$$r = \frac{432 - \frac{2880}{6}}{\sqrt{\left[444 - \frac{2304}{6}\right]\left[646 - \frac{3600}{6}\right]}}$$

$$r = \frac{432 - 480}{\sqrt{[444 - 384][646 - 600]}}$$

$$r = \frac{-48}{\sqrt{[60][46]}}$$

$$r = \frac{-48}{\sqrt{2760}}$$

$$r = \frac{-48}{52.5357}$$

$$r = -0.91$$

Are you starting to get the hang of it? How close did your estimate come to our r of –0.91? If your answer was one one-hundredth (.01) off, that was due to rounding and can be ignored. The way to really learn correlation is by doing a lot of problems, and that's just what we'll be doing for the rest of the chapter.

Use the data in Table 10.6 to find the correlation between X and Y. Then check my results.

Table 10.6

X	X²	Y	Y²	XY
20		1		
17		4		
13		9		
10		7		
8		9		
4		6		

Solution:

Table 10.7

X	X²	Y	Y²	XY
20	400	1	1	20
17	289	4	16	68
13	169	9	81	127
10	100	7	49	70
8	64	9	81	72
4	16	6	36	24
72	1038	36	264	381
$\overline{X} = 12$		$\overline{Y} = 6$		

$$r = \frac{\sum XY - \frac{(\sum X)(\sum Y)}{n}}{\sqrt{\left[\sum X^2 - \frac{(\sum X)^2}{n}\right]\left[\sum Y^2 - \frac{(\sum Y)^2}{n}\right]}}$$

$$r = \frac{381 - \frac{(72)(36)}{6}}{\sqrt{\left[1038 - \frac{(72)^2}{6}\right]\left[264 - \frac{(36)^2}{6}\right]}}$$

$$r = \frac{381 - \frac{2592}{6}}{\sqrt{\left[1038 - \frac{5184}{6}\right]\left[264 - \frac{1296}{6}\right]}}$$

$$r = \frac{381 - 432}{\sqrt{[1038 - 864][264 - 216]}}$$

$$r = \frac{-51}{\sqrt{[174][48]}}$$

$$r = \frac{-51}{\sqrt{8352}}$$

$$r = \frac{-51}{91.3893}$$

$$r = -0.56$$

How did the r you calculated compare with your estimate? Not only should we expect an r that is negative, but we should also notice that as X declines, Y does not rise consistently. So we have a rather weak negative correlation.

So far I've set up all the problems in a table and you carried out all the calculations very routinely. But out there in the real world, no one will do that for you. So no more Mr. Nice Guy. Now you'll have to set up your own problems.

Let's return to the question of how closely height and weight are correlated. Suppose we've made these six observations: (5'6", 150 lbs.), (5'8", 170 lbs.), (5'9", 160 lbs.), (5'11", 200), (6'0", 240), and (6'2", 220). Fill in these observations in the X and Y columns of Table 10.8. Be sure to convert your heights into inches. And does it matter whether you make height or weight X?

It doesn't matter. Height could be X and weight could be Y. Or vice versa. After you've solved for r, check your results against mine.

Table 10.8

X	X²	Y	Y²	XY

Solution:

Table 10.9

X	X²	Y	Y²	XY
66	4,356	150	22,500	9,900
68	4,624	170	28,900	11,560
69	4,761	160	25,600	11,040
71	5,041	200	40,000	14,200
72	5,184	240	57,600	17,280
74	5,476	220	48,400	16,280
420	29,442	1,140	223,000	80,260
$\overline{X} = 7$		$\overline{Y} = 190$		

$$r = \frac{\sum XY - \frac{(\sum X)(\sum Y)}{n}}{\sqrt{\left[\sum X^2 - \frac{(\sum X)^2}{n}\right]\left[\sum Y^2 - \frac{(\sum Y)^2}{n}\right]}}$$

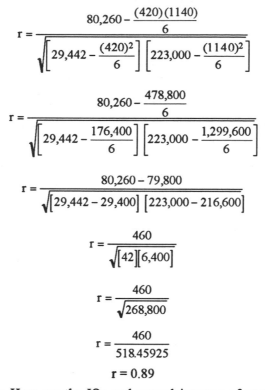

$$r = \frac{80,260 - \frac{(420)(1140)}{6}}{\sqrt{\left[29,442 - \frac{(420)^2}{6}\right]\left[223,000 - \frac{(1140)^2}{6}\right]}}$$

$$r = \frac{80,260 - \frac{478,800}{6}}{\sqrt{\left[29,442 - \frac{176,400}{6}\right]\left[223,000 - \frac{1,299,600}{6}\right]}}$$

$$r = \frac{80,260 - 79,800}{\sqrt{\left[29,442 - 29,400\right]\left[223,000 - 216,600\right]}}$$

$$r = \frac{460}{\sqrt{\left[42\right]\left[6,400\right]}}$$

$$r = \frac{460}{\sqrt{268,800}}$$

$$r = \frac{460}{518.45925}$$

$$r = 0.89$$

Problem: Here are the IQs and annual incomes of seven people: (93, $14,000); (98, $17,000); (100, $23,000); (102, $20,000); (109, $25,000); (119, $80,000); (128, $66,000). Use this information to fill in Table 10.10. (You can get rid of sets of three zeroes following the thousands of dollars of income.) Then compute the correlation between IQ and annual income, and compare your work with mine.

Table 10.10

X	X²	Y	Y²	XY

Solution:

Table 10.11

X	X²	Y	Y²	XY
93	8,649	14	196	1,302
98	9,604	17	289	1,666
100	10,000	23	529	2,300
102	10,404	20	400	2,040
109	11,881	25	625	2,725
119	14,161	80	6,400	9,520
128	16,384	66	4,356	8,448
749	81,083	245	12,795	28,001
$\overline{X} = 107$		$\overline{Y} = 35$		

$$r = \frac{\sum XY - \frac{(\sum X)(\sum Y)}{n}}{\sqrt{\left[\sum X^2 - \frac{(\sum X)^2}{n}\right]\left[\sum Y^2 - \frac{(\sum Y)^2}{n}\right]}}$$

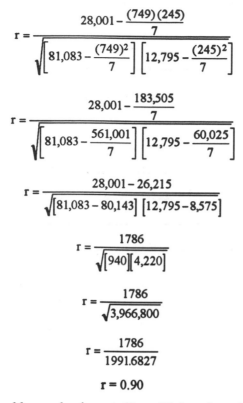

$$r = \frac{28{,}001 - \dfrac{(749)(245)}{7}}{\sqrt{\left[81{,}083 - \dfrac{(749)^2}{7}\right]\left[12{,}795 - \dfrac{(245)^2}{7}\right]}}$$

$$r = \frac{28{,}001 - \dfrac{183{,}505}{7}}{\sqrt{\left[81{,}083 - \dfrac{561{,}001}{7}\right]\left[12{,}795 - \dfrac{60{,}025}{7}\right]}}$$

$$r = \frac{28{,}001 - 26{,}215}{\sqrt{\left[81{,}083 - 80{,}143\right]\left[12{,}795 - 8{,}575\right]}}$$

$$r = \frac{1786}{\sqrt{[940][4{,}220]}}$$

$$r = \frac{1786}{\sqrt{3{,}966{,}800}}$$

$$r = \frac{1786}{1991.6827}$$

$$r = 0.90$$

One more problem and we're out of here. We have here the average number of packs of cigarettes a day these eight people smoked and their age at death: (1.4, 77) (2.0, 83), (2.7, 74), (3.0, 72), (3.2, 63), (3.6, 66) (4.2, 64), and (4.7, 61). Please fill in Table 10.12, compute r, and then compare your work with mine.

Table 10.12

X	X²	Y	Y²	XY

Solution:

Table 10.13

X	X²	Y	Y²	XY
1.4	1.96	77	5,929	150.92
2.0	4.00	83	6,889	166.00
2.7	7.29	74	5,476	199.80
3.0	9.00	72	5,184	216.00
3.2	10.24	63	3,969	201.60
3.6	12.96	66	4,356	237.60
4.2	17.64	64	4,096	268.80
4.7	22.09	61	3,721	286.70
24.8	85.18	560	39,620	1,727.42

$$\overline{X} = 31 \qquad\qquad \overline{Y} = 70$$

$$r = \frac{\Sigma XY - \frac{(\Sigma X)(\Sigma Y)}{n}}{\sqrt{\left[\Sigma X^2 - \frac{(\Sigma X)^2}{n}\right]\left[\Sigma Y^2 - \frac{(\Sigma Y)^2}{n}\right]}}$$

$$r = \frac{1727.42 - \frac{(24.8)(560)}{8}}{\sqrt{\left[85.18 - \frac{(24.8)^2}{8}\right]\left[39,620 - \frac{(560)^2}{8}\right]}}$$

$$r = \frac{1727.42 - \frac{13,888}{8}}{\sqrt{\left[85.18 - \frac{615.04}{8}\right]\left[39,620 - \frac{313,600}{8}\right]}}$$

$$r = \frac{1727.42 - 1736}{\sqrt{[85.18 - 76.88][39,620 - 39,200]}}$$

$$r = \frac{-8.58}{\sqrt{[8.3][420]}}$$

$$r = \frac{-8.58}{\sqrt{3486}}$$

$$r = \frac{-8.58}{59.0424}$$

$$r = -0.15$$

Chapter 11

Prediction and Regression

Every college admission officer knows that applicants with high SAT scores tend to have higher college grade point averages than applicants with low SAT scores. But how *much* higher? By the time you've worked your way through this chapter you'll be able to make forecasts like this yourself.

First we'll be predicting scores on an individual basis. For instance, if a baby is born weighing seven pounds, what will her predicted weight be on her first birthday?

What we can do on an individual basis, we can also do on a larger scale. We call this regression analysis, which we'll be doing in the second part of this chapter.

Predicting Individual Scores

Let's see how we can predict a baby's weight on her first birthday if all we know is that she weighed seven pounds at birth. Suppose previous studies have indicated a correlation of 0.0 between baby girls' birth weights and their weights on their first birthdays. Can we still make a prediction?

The answer is yes, but it won't be anything to write home about. The best we can do is predict that she will weigh the average or mean weight of one–year–old girls.

But what if there happens to be a correlation between birth weight and weight at one year? Then we're in business. We can use this equation to predict her weight:

$$Y^1 = \overline{Y} + r\frac{S_y}{S_x}(X - \overline{X})$$

Of course this equation isn't much help unless you know what each of the letters stands for. Y^1 is her predicted weight on her first birthday. That's the unknown we need to find.

183

$\overline{Y}$ is the mean weight at age one; r is the coefficient of correlation; s_y is the standard deviation of weights at age one; s_x is the standard deviation of birth weights; X is the birth weight; and $\overline{X}$ is the mean birth weight.

To solve this problem you need to know each of the variables on the right side of the equation. Then, just plug these numbers into the equation and solve for Y^1.

Given: Y = 21; r = 0.6; s_y = 6; s_x = 1; and $\overline{X}$ = 6.5. By the way, how much is X? X is the birth weight of 7.

Now just substitute these numbers for the letters in the equation and solve for Y^1. Then check your work with mine:

Solution:
$$Y^1 = \overline{Y} + r\frac{S_y}{S_x}(X - \overline{X})$$

$$Y^1 = 21 + 0.6 \times \left(\frac{6}{1}\right)(7 - 6.5)$$

$$Y^1 = 21 + 3.6\,(0.5)$$
$$Y^1 = 21 + 1.8$$
$$Y^1 = 22.8$$

Let's make a couple of observations. If r had been zero, then Y^1 would have been equal to $\overline{Y}$. Similarly, if X had equaled $\overline{X}$, then Y^1 would have been equal to $\overline{Y}$. In both cases, then, all the terms on the right side of the equation beyond $\overline{Y}$ would have been equal to zero.

Here's another problem: Suppose IQ is used as a predictor of average lifetime annual income. If Mahmoud El–Kati has an IQ of 120, there is a correlation of 0.8 between IQ and average lifetime annual income, the standard deviation of average lifetime annual income is 20, the standard deviation of IQ is 16, the mean IQ is 100, and the average lifetime income is $25,000, find Mr. El–Kati's average lifetime expected income. Before you start, ask yourself one basic question: Which are the X's and which are the Y's?

The X's are the IQs and the Y's are the incomes. Now you're ready to substitute numbers into the equation and solve for Y. After you've done so, check your work.

Solution:
$$Y^1 = \overline{Y} + r\frac{S_y}{S_x}(X - \overline{X})$$

$$Y^1 = \$25,000 + 0.8\left(\frac{20}{16}\right)(120 - 100)$$

$$Y^1 = \$25,000 + 0.8\,(1.25)\,(20)$$
$$Y^1 = \$25,000 + 1.0\,(20)$$

$$Y^1 = \$25,000 + \$20,000$$
$$Y^1 = \$45,000$$

Here's a similar problem. Alice Maxwell has an IQ of 90. Find her average lifetime annual income using the same data we used in the previous problem.

Solution:
$$Y^1 = \overline{Y} + r\frac{S_y}{S_x}(X - \overline{X})$$

$$Y^1 = \$25,000 + 0.8\left(\frac{20}{16}\right)(90 - 100)$$

$$Y^1 = \$25,000 + 1\,(-10)$$
$$Y^1 = \$25,000 - \$10,000$$
$$Y^1 = \$15,000$$

When we predicted weight at age one based on birth weight, we were predicting the future. But as the term "predict" is used in statistics, we may mean simply that we are using information about one variable to make an informed guess, or estimate, of another. For instance, we might have used average lifetime annual income to "predict" IQ.

Problem: A man is 6' tall. Predict his weight if the average weight for men in his age group is 160 pounds, the coefficient of correlation is 0.7, the standard deviation of weight is 20, the standard deviation of height is 4, and the average height is 5'9".

Solution:
$$Y^1 = \overline{Y} + r\frac{S_y}{S_x}(X - \overline{X})$$

$$Y^1 = 160 + 0.7\left(\frac{20}{4}\right)(72 - 69)$$

$$Y^1 = 160 + 0.7(5)(3)$$
$$Y^1 = 160 + 0.7(15)$$
$$Y^1 = 160 + 10.5$$
$$Y^1 = 170.5 \text{ pounds}$$

This next problem is a little tricky. Harrison Buford has gotten 20 hours of tutoring in math during the two weeks before his math exam, while the mean number of hours of tutoring of students in his class was 8 hours. Predict his score on the exam if the mean score was 78, the coefficient of correlation is –0.3, the standard deviation of math scores was 10, and the standard deviation of hours of tutoring was 5.

Solution: $$Y^1 = \overline{Y} + r\frac{S_y}{S_x}(X - \overline{X})$$

$$Y^1 = 78 + (-0.3)\left(\frac{10}{5}\right)(20 - 8)$$

$$Y^1 = 78 - (0.3)\,(2)\,(12)$$

$$Y^1 = 78 - (0.6)\,(12)$$

$$Y^1 = 78 - 7.2$$

$$Y^1 = 70.8$$

Here's one last problem. George Goldberg is 50 years old and has a cholesterol level of 180. The average cholesterol level for men his age is 200. If the average life expectancy for men 50 years old is 83, what is Mr. Goldberg's life expectancy if the coefficient of correlation is – 0.8, the standard deviation of life expectancy is 8, and the standard deviation of the cholesterol level is 20?

Solution: $$Y^1 = \overline{Y} + r\frac{S_y}{S_x}(X - \overline{X})$$

$$Y^1 = 83 - 0.8\left(\frac{8}{20}\right)(180 - 200)$$

$$Y^1 = 83 - 0.8\,(.4)\,(-20)$$

$$Y^1 = 83 - 0.32\,(-20)$$

$$Y^1 = 83 - (-6.4)$$

$$Y^1 = 83 + 6.4$$

$$Y^1 = 89.4$$

Freehand Regression Line Fitting

You may recall (but probably won't) that back in Chapter 7 (Standard Deviation) we said that the mean was a point in a distribution that makes the sum of the squares of deviations from it minimal. In other words, the least squares. When applying the least squares method to correlation and regression, the line of best fit is defined as the line that minimizes the squared deviations around it. We call that line a *regression line*.

Let's redraw Figure 10.1d and 10.1e from the last chapter as Figures 11.1a and 11.1b. Figure 11.1a illustrates a perfectly positive correlation, and Figure 11.1b illustrates a perfectly negative correlation.

Figure 11.1a: Perfectly Positive Correlation

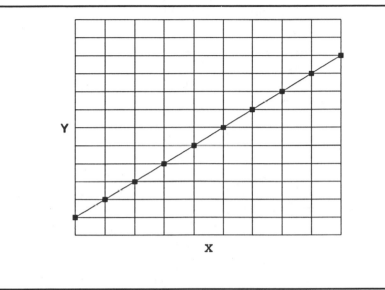

Figure 11.1b: Perfectly Negative Correlation

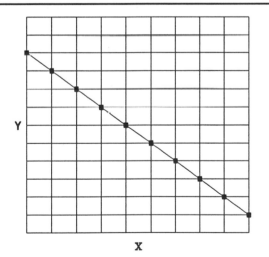

Now let's redraw Figure 10.1b as Figure 11.2a. Would it be possible to draw a regression line through the points on this scatter diagram? It would not only be possible, but it is actually done in Figure 11.2b.

Figure 11.2a: Hypothetical Sampling of Grade Point Average and Hours Per Day Spent Watching TV of 250 College Students

---

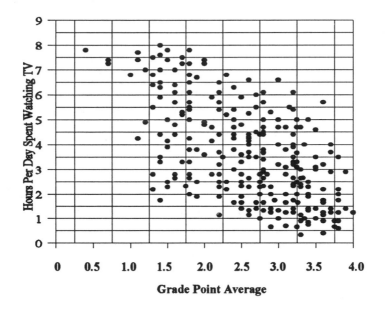

Grade Point Average

Figure 11.2b: Hypothetical Sampling of Grade Point Average and Hours Per Day Spent Watching TV of 250 College Students

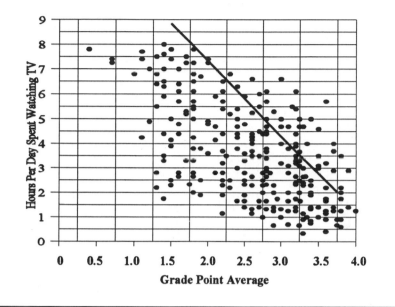

And now please consider Figure 11.3a, which is a duplicate of Figure 10.1a from the last chapter. See if you can draw a regression line on this scatter diagram.

Figure 11.3a: Hypothetical Sampling of the Heights and Weights of 260 Men, Aged 21–59

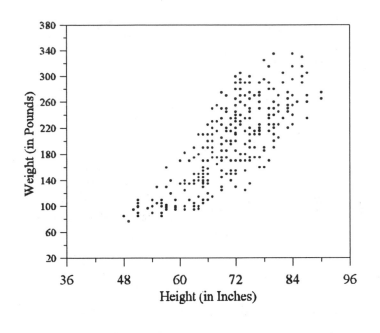

Compare your line with mine in Figure 11.3b. Freehand line fitting is really an art, so as long as your line is straight and slopes from the upper left to the lower right, then you're probably right on the money.

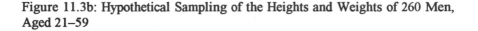

Figure 11.3b: Hypothetical Sampling of the Heights and Weights of 260 Men, Aged 21–59

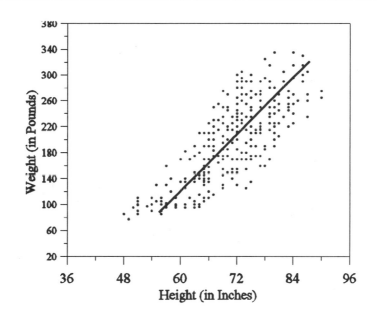

Constructing Regression Lines by Predicting Y-Scores

You may be asking yourself this question: Self, is there a more precise way of drawing these regression lines? Funny you should ask, because it just so happens there is.

Do you remember how we predicted the Y-score, called Y^1? To construct a regression line predicting Y^1 from X, all we need to do is take two extreme values of X, predict Y^1 from each of these values, and then join these two points on a scatter diagram.

Let's do that now using the data we obtained in the IQ and average lifetime annual income prediction problems we solved near the beginning of the chapter. We found that a person with an IQ of 120 had a predicted average lifetime annual income of $45,000. And someone with a 90 IQ had a predicted average lifetime annual income of $15,000. So we've got the two points we need to plot our regression line.

A point on a graph is located by the coordinates (X,Y). Our two points, then, are located by the coordinates (120, $45,000) and (90, $15,000). As you probably noticed, our X axis is the horizontal axis and the Y axis is the vertical axis. Please plot these two points (or coordinates) in Figure 11.4a. Then use a ruler or straightedge to connect them with a straight line.

Figure 11.4a

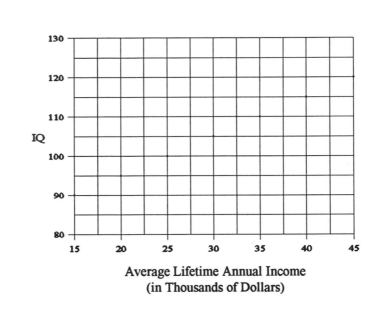

Average Lifetime Annual Income
(in Thousands of Dollars)

You should have gotten a straight line sloping from the lower left to the upper right like the one shown in Figure 11.4b.

Figure 11.4b

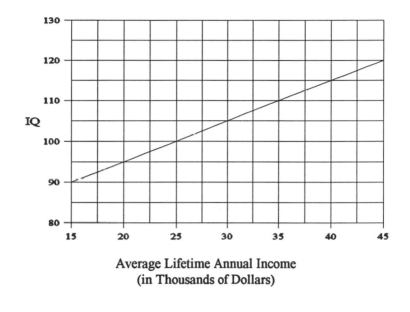

IQ

Average Lifetime Annual Income
(in Thousands of Dollars)

Now we'll use some data from the next problem to help us plot a regression line in Figure 11.5a. We saw that a man who was 6 feet tall had a predicted weight of 170.5 pounds. Find the predicted weight of a man who is 5 feet tall. Then use the pair of coordinates (the heights and weights of both men) to draw the regression line.

Figure 11.5a

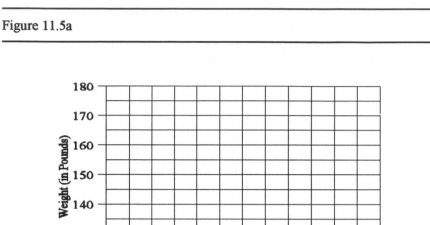

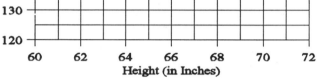

Solution:

$$Y^1 = \overline{Y} + r\frac{S_y}{S_x}(X - \overline{X})$$

$$Y^1 = 160 + (0.7)\left(\frac{20}{4}\right)(60 - 69)$$

$$Y^1 = 160 + (0.7)\,(5)\,(-9)$$

$$Y^1 = 160 + (3.5)\,(-9)$$

$$Y^1 = 160 + (-31.5)$$

$$Y^1 = 160 - 31.5$$

$$Y^1 = 128.5 \text{ pounds}$$

The two coordinates are (60, 128.5) and (72, 170.5). These are used to plot the regression line shown in Figure 11.5b.

Figure 11.5b

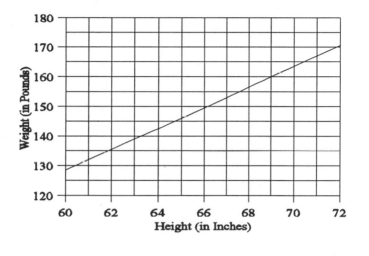

Let's use the data from the next problem to plot one more regression line. In that problem we found that Harrison Buford, who had gotten 20 hours of math tutoring, had a predicted grade of 70.8. Suppose that Yogesh Gandhi did not get any tutoring. Find his predicted score and then plot your regression line in Figure 11.6a.

Figure 11.6a

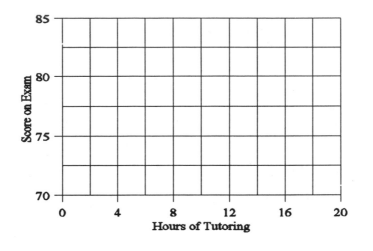

Solution:

$$Y^1 = \overline{Y} + r\frac{S_y}{S_x}(X - \overline{X})$$

$$Y^1 = 78 + (-0.3)\left(\frac{10}{5}\right)(0-8)$$

$$Y^1 = 78 + (-0.3)\,(2)\,(-8)$$

$$Y^1 = 78 + (-0.6)\,(-8)$$

$$Y^1 = 78 + (+4.8)$$

$$Y^1 = 78 + 4.8$$

$$Y^1 = 82.8$$

Our plot coordinates are (0, 82.8) and (20, 70.8). Using them we plot our regression line in Figure 11.6b.

Figure 11.6b

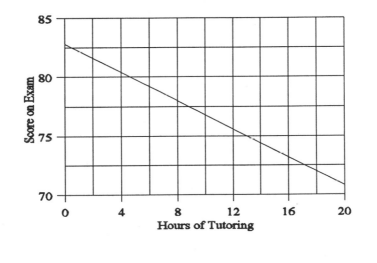

Using Regression Lines

Now that we've drawn all these regression lines, how can we use them? Good question. We can use them to predict scores without having to write down that formula and do all of those calculations. All we need to do is to look at the regression line.

For instance, if you look at Figure 11.6b, you can see instantly that if you had six hours of tutoring, you would have a predicted exam score of 79. What would your exam score be if you had 16 hours of tutoring?

You would have a score of 73. Moving back to Figure 11.5b, what would be your predicted weight if you were 5'4"?

It would be 142.5 pounds. One last question. If you were 5'7", what would your predicted weight be?

It would be 152.5 pounds, or possibly as much as 153. That's it! No more questions!

Appendix

Table A

Proportions of Area under the Normal Curve

(A) z	(B) Area between Mean and z	(C) Area beyond z	(A) z	(B) Area between Mean and z	(C) Area beyond z	(A) z	(B) Area between Mean and z	(C) Area beyond z
0.00	.0000	.5000	1.12	.3686	.1314	2.24	.4875	.0125
0.01	.0040	.4960	1.13	.3708	.1292	2.25	.4878	.0122
0.02	.0080	.4920	1.14	.3729	.1271	2.26	.4881	.0119
0.03	.0120	.4880	1.15	.3749	.1251	2.27	.4884	.0116
0.04	.0160	.4840	1.16	.3770	.1230	2.28	.4887	.0113
0.05	.0199	.4801	1.17	.3790	.1210	2.29	.4890	.0110
0.06	.0239	.4761	1.18	.3810	.1190	2.30	.4893	.0107
0.07	.0279	.4721	1.19	.3830	.1170	2.31	.4896	.0104
0.08	.0319	.4681	1.20	.3849	.1151	2.32	.4898	.0102
0.09	.0359	.4641	1.21	.3869	.1131	2.33	.4901	.0099
0.10	.0398	.4602	1.22	.3888	.1112	2.34	.4904	.0096
0.11	.0438	.4562	1.23	.3907	.1093	2.35	.4906	.0094
0.12	.0478	.4522	1.24	.3925	.1075	2.36	.4909	.0091
0.13	.0517	.4483	1.25	.3944	.1056	2.37	.4911	.0089
0.14	.0557	.4443	1.26	.3962	.1038	2.38	.4913	.0087
0.15	.0596	.4404	1.27	.3980	.1020	2.39	.4916	.0084
0.16	.0636	.4364	1.28	.3997	.1003	2.40	.4918	.0082
0.17	.0675	.4325	1.29	.4015	.0985	2.41	.4920	.0080

(A) z	(B) Area between Mean and z	(C) Area beyond z	(A) z	(B) Area between Mean and z	(C) Area beyond z	(A) z	(B) Area between Mean and z	(C) Area beyond z
0.18	.0714	.4286	1.30	.4032	.0968	2.42	.4922	.0078
0.19	.0753	.4247	1.31	.4049	.0951	2.43	.4925	.0075
0.20	.0793	.4207	1.32	.4066	.0934	2.44	.4927	.0073
0.21	.0832	.4168	1.33	.4082	.0918	2.45	.4929	.0071
0.22	.0871	.4129	1.34	.4099	.0901	2.46	.4931	.0069
0.23	.0910	.4090	1.35	.4115	.0885	2.47	.4932	.0068
0.24	.0948	.4052	1.36	.4131	.0869	2.48	.4934	.0066
0.25	.0987	.4013	1.37	.4147	.0853	2.49	.4936	.0064
0.26	.1026	.3974	1.38	.4162	.0838	2.50	.4938	.0062
0.27	.1064	.3936	1.39	.4177	.0823	2.51	.4940	.0060
0.28	.1103	.3897	1.40	.4192	.0808	2.52	.4941	.0059
0.29	.1141	.3859	1.41	.4207	.0793	2.53	.4943	.0057
0.30	.1179	.3821	1.42	.4222	.0778	2.54	.4945	.0055
0.31	.1217	.3783	1.43	.4236	.0764	2.55	.4946	.0054
0.32	.1255	.3745	1.44	.4251	.0749	2.56	.4948	.0052
0.33	.1293	.3707	1.45	.4265	.0735	2.57	.4949	.0051
0.34	.1331	.3669	1.46	.4279	.0721	2.58	.4951	.0049
0.35	.1368	.3632	1.47	.4292	.0708	2.59	.4952	.0048
0.36	.1406	.3594	1.48	.4306	.0694	2.60	.4953	.0047
0.37	.1443	.3557	1.49	.4319	.0681	2.61	.4955	.0045
0.38	.1480	.3520	1.50	.4332	.0668	2.62	.4956	.0044
0.39	.1517	.3483	1.51	.4345	.0655	2.63	.4957	.0043
0.40	.1554	.3446	1.52	.4357	.0643	2.64	.4959	.0041
0.41	.1591	.3409	1.53	.4370	.0630	2.65	.4960	.0040
0.42	.1628	.3372	1.54	.4382	.0618	2.66	.4961	.0039
0.43	.1664	.3336	1.55	.4394	.0606	2.67	.4962	.0038
0.44	.1700	.3300	1.56	.4406	.0594	2.68	.4963	.0037
0.45	.1736	.3264	1.57	.4418	.0582	2.69	.4964	.0036
0.46	.1772	.3228	1.58	.4429	.0571	2.70	.4965	.0035
0.47	.1808	.3192	1.59	.4441	.0559	2.71	.4966	.0034
0.48	.1844	.3156	1.60	.4452	.0548	2.72	.4967	.0033

(A) z	(B) Area between Mean and z	(C) Area beyond z	(A) z	(B) Area between Mean and z	(C) Area beyond z	(A) z	(B) Area between Mean and z	(C) Area beyond z
0.49	.1879	.3121	1.61	.4463	.0537	2.73	.4968	.0032
0.50	.1915	.3085	1.62	.4474	.0526	2.74	.4969	0031
0.51	.1950	.3050	1.63	.4484	.0516	2.75	.4970	.0030
0.52	.1985	.3015	1.64	.4495	.0505	2.76	.4971	.0029
0.53	.2019	.2981	1.65	.4505	.0495	2.77	.4972	.0028
0.54	.2054	.2946	1.66	.4515	.0485	2.78	.4973	.0027
0.55	.2088	.2912	1.67	.4525	.0475	2.79	.4974	.0026
0.56	.2123	.2877	1.68	.4535	.0465	2.80	.4974	.0026
0.57	.2157	.2843	1.69	.4545	.0455	2.81	.4975	.0025
.058	.2190	.2180	1.70	.4554	.0446	2.82	.4976	.0024
.059	.2224	.2776	1.71	.4564	.0436	2.83	.4977	.0023
0.60	.2257	.2743	1.72	.4573	.0427	2.84	.4977	.0023
0.61	.2291	.2709	1.73	.4582	.0418	2.85	.4978	.0022
0.62	.2324	.2676	1.74	.4591	.0409	2.86	.4979	.0021
0.63	.2357	.2643	1.75	.4599	.0401	2.87	.4979	.0021
0.64	.2389	.2611	1.76	.4608	.0392	2.88	.4980	.0020
0.65	.2422	.2578	1.77	.4616	.0384	2.89	.4981	.0019
0.66	.2454	.2546	1.78	.4625	.0375	2.90	.4981	.0019
0.67	.2486	.2514	1.79	.4633	.0367	2.91	.4982	.0018
0.68	.2517	.2483	1.80	.4641	.0359	2.92	.4982	.0018
0.69	.2549	.2451	1.81	.4649	.0351	2.93	.4983	.0017
0.70	.2580	.2420	1.82	.4656	.0344	2.94	.4984	.0016
0.71	.2611	.2389	1.83	.4664	.0336	2.95	.4984	.0016
0.72	.2642	.2358	1.84	.4671	.0329	2.96	.4985	.0015
0.73	.2673	.2327	1.85	.4678	.0322	2.97	.4985	.0015
0.74	.2704	.2296	1.86	.4686	.0314	2.98	.4986	.0014
0.75	.2734	.2266	1.87	.4693	.0307	2.99	.4986	.0014
0.76	.2764	.2236	1.88	.4699	.0301	3.00	.4987	.0013
0.77	.2794	.2206	1.89	.4706	.0294	3.01	.4987	.0013
0.78	.2823	.2177	1.90	.4713	.0287	3.02	.4987	.0013
0.79	.2852	.2148	1.91	.4719	.0281	3.03	.4988	.0012

(A) z	(B) Area between Mean and z	(C) Area beyond z	(A) z	(B) Area between Mean and z	(C) Area beyond z	(A) z	(B) Area between Mean and z	(C) Area beyond z
0.80	.2881	.2119	1.92	.4726	.0274	3.04	.4988	.0012
0.81	.2910	.2090	1.93	.4732	.0268	3.05	.4989	.0011
0.82	.2939	.2061	1.94	.4738	.0262	3.06	.4989	.0011
0.83	.2967	.2033	1.95	.4744	.0256	3.07	.4989	.0011
0.84	.2995	.2005	1.96	.4750	.0250	3.08	.4990	.0010
0.85	.3023	.1977	1.97	.4756	.0244	3.09	.4990	.0010
0.86	.3051	.1949	1.98	.4761	.0239	3.10	.4990	.0010
0.87	.3078	.1922	1.99	.4767	.0233	3.11	.4991	.0009
0.88	.3106	.1894	2.00	.4772	.0228	3.12	.4991	.0009
0.89	.3133	.1867	2.01	.4778	.0222	3.13	.4991	.0009
0.90	.3159	.1841	2.02	.4783	.0217	3.14	.4992	.0008
0.91	.3186	.1814	2.03	.4788	.0212	3.15	.4992	.0008
0.92	.3212	.1788	2.04	.4793	.0207	3.16	.4992	.0008
0.93	.3238	.1762	2.05	.4798	.0202	3.17	.4992	.0008
0.94	.3264	.1736	2.06	.4803	.0197	3.18	.4993	.0007
0.95	.3289	.1711	2.07	.4808	.0192	3.19	.4993	.0007
0.96	.3315	.1685	2.08	.4812	.0188	3.20	.4993	.0007
0.97	.3340	.1660	2.09	.4817	.0183	3.21	.4993	.0007
0.98	.3365	.1635	2.10	.4821	.0179	3.22	.4994	.0006
0.99	.3389	.1611	2.11	.4826	.0174	3.23	.4994	.0006
1.00	.3413	.1587	2.12	.4830	.0170	3.24	.4994	.0006
1.01	.3438	.1562	2.13	.4834	.0166	3.25	.4994	.0006
1.02	.3461	.1539	2.14	.4838	.0162	3.30	.4995	.0005
1.03	.3485	.1515	2.15	.4842	.0158	3.35	.4996	.0004
1.04	.3508	.1492	2.16	.4846	.0154	3.40	.4997	.0003
1.05	.3531	.1469	2.17	.4850	.0150	3.45	.4997	.0003
1.06	.3554	.1446	2.18	.4854	.0146	3.50	.4998	.0002
1.07	.3577	.1423	2.19	.4857	.0143	3.60	.4998	.0002
1.08	.3599	.1401	2.20	.4861	.0139	3.70	.4999	.0001
1.09	.3621	.1379	2.21	.4864	.0136	3.80	.4999	.0001

(A) z	(B) Area between Mean and z	(C) Area beyond z	(A) z	(B) Area between Mean and z	(C) Area beyond z	(A) z	(B) Area between Mean and z	(C) Area beyond z
1.10	.3643	.1357	2.22	.4868	.0132	3.90	.49995	.00005
1.11	.3665	.1335	2.23	.4871	.0129	4.00	.49997	.00003
0.17	.0675	.4325	1.29	.4015	.0985	2.41	.4920	.0080

Recommendations for Further Reading

Remember the title of this book? *Chances Are* this is the only statistics book you'll ever need! But what if you *do* need to go on studying statistics?

I would recommend four books: Susan H. Gray, *No-Frills Statistics: A Guide for the First-Year Student* (Littlefield, Adams); Donald Koosis, *Statistics* (John Wiley and Sons); Murray Spiegel, *Statistics* (Shaum); and David Voekler and Peter Orton, *Statistics* (Cliff Quick Review). Also, Doubleday, Barron's, HBJ, and REA (Research and Education Associates) all publish two or three statistics review books in their college outline series.

At the end of Chapter 1 I used some illustrations from *How to Lie with Statistics*, a best-seller written by Darrell Huff (Norton). This nontechnical book does a marvelous job explaining just what its title promises, especially with respect to graphs.

Even if you decide not to go on with statistics, you'll need to use what you've learned in this book to ensure that your number-crunching skills don't grow rusty. As you've heard so many times before: use it or lose it.

About the Author

Steve Slavin received his B.A. in economics from Brooklyn College and his Ph.D. in economics from New York University. For 31 years he taught at New York Institute of Technology, Brooklyn College, St. Francis College (Brooklyn), Union County College (New Jersey), and the New School for Social Research. He is now a full-time writer.

He has written ten other books: *The Einstein Syndrome: Corporate Anti-Semitism in America Today* (University Press of America); *Jelly Bean Economics: Reaganomics in the Early 1980s* (Philosophical Library); *Economics: A Self-Teaching Guide* (John Wiley and Sons); and *Economics* (an introductory college text used in over 150 schools, now in its fifth edition, published by McGraw-Hill). Also, *All the Math You'll Ever Need; Practical Algebra; Quick Algebra Review; Math for Your 1st and 2nd Grader; Quick Business Math* (all published by Wiley); and *Everyday Math in 20 Minutes a Day* (LearningExpress).